Bioactive Compounds Targeting Neurodegenerative Diseases

Authored by

Shivendra Mani Tripathi

Faculty of Pharmaceutical Sciences
Mahayogi Gorakhnath University Gorakhpur
Uttar Pradesh
India

Sudhanshu Mishra

Faculty of Pharmaceutical Sciences
Mahayogi Gorakhnath University Gorakhpur
Uttar Pradesh
India

Rishabha Malviya

Department of Pharmacy
School of Medical and Allied Sciences
Galgotias University, Greater Noida
Uttar Pradesh, India

&

Smriti Ojha

Department of Pharmaceutical Science
& Technology, Madan Mohan Malaviya
University of Technology, Gorakhpur
Uttar Pradesh, India

Bioactive Compounds Targeting Neurodegenerative Diseases

Authors: Shivendra Mani Tripathi, Sudhanshu Mishra, Rishabha Malviya, Smriti Ojha

ISBN (Online): 978-981-5324-84-6

ISBN (Print): 978-981-5324-85-3

ISBN (Paperback): 978-981-5324-86-0

need for a court order if at any point you breach any terms of this License Agreement. In no event will any delay or failure by Bentham Science Publishers in enforcing your compliance with this License Agreement constitute a waiver of any of its rights.

3. You acknowledge that you have read this License Agreement, and agree to be bound by its terms and conditions. To the extent that any other terms and conditions presented on any website of Bentham Science Publishers conflict with, or are inconsistent with, the terms and conditions set out in this License Agreement, you acknowledge that the terms and conditions set out in this License Agreement shall prevail.

Bentham Science Publishers Pte. Ltd.
80 Robinson Road #02-00
Singapore 068898
Singapore
Email: subscriptions@benthamscience.net

BENTHAM SCIENCE

CONTENTS

FOREWORD

This book is an invaluable resource that presents information about neurodegeneration, pathophysiology, and the management aspects of "Bioactive compounds for neurodegenerative diseases", authored by Mr. Sudhanshu Mishra and his team. The book chapters, which were written by a varied group of experts in the field, researchers, and professionals, cover a wide range of subjects, including various approaches to disease management, natural biomolecules, and their role in managing disease progression.

Students, researchers, and neurologists who are interested in the subject matter are the intended audience for the book. The book inspires readers to indulge in the knowledge about neurodegeneration, participate in conversations, and spread the word.

Bringing out a current peer-reviewed collection under the title "Bioactive compounds for neurodegenerative diseases " with contributors spanning across the world has been expertly done by the editors. It is quite amazing to see how the editors have attempted to encompass such a broad and dynamic subject in several engrossing pieces that shed light on the role of bioactive compounds in disease management. Enthusiastic readers will be captivated by the appealing pictures that illustrate intricate theoretical and experimental aspects as well as the overall production design. This book's goal is to document and explore advancements in both the diagnosis and possible treatment of NDDs using tried-and-true methods. It brings together scientists from all over the world with varying specialties and areas of interest in science to concentrate on NDDs.

It gives me great pleasure to provide the foreword for this esteemed, multi-authored, internationally peer-reviewed publication on a subject extremely important for the well-being of all. I hope the editors have a prosperous future and would love to see this modest start become a regular series. My heartfelt gratitude goes out to every contributor for their outstanding work in making this book a success.

With best wishes

Rohit Kumar Verma
International Medical University, Malaysia

PREFACE

We are thrilled to introduce our readers to the book "Bioactive Compounds Targeting Neurodegenerative Diseases." This comprehensive work is the culmination of extensive research efforts aimed at providing researchers and healthcare professionals with a valuable resource to address the challenges posed by neurodegenerative diseases.

Neurodegenerative diseases have long been a significant concern, and as our understanding of these conditions evolves, so does our approach to combating them. This book aims to bridge the gap between knowledge and action by empowering readers with the information they need to effectively manage and combat neurodegenerative diseases.

Each of the upcoming chapters delves into different aspects of neurodegenerative diseases, from the underlying mechanisms to the latest advancements in treatment strategies along with the challenges and intellectual properties. With the insights provided in this book, readers will be better equipped to engage in informed discussions with healthcare providers, make sound decisions, and take proactive steps toward managing neurodegenerative diseases.

This book is not only intended for researchers and healthcare professionals but also for caregivers, patients, and anyone seeking to understand and support those affected by neurodegenerative diseases. By fostering constructive dialogue and collaboration, this book catalyzes collective efforts to combat these challenging conditions. We trust that this book will serve as a trusted companion in the quest for improved treatments and outcomes for neurodegenerative diseases, inspiring readers to play an active role in promoting brain health and well-being.

Shivendra Mani Tripathi
Shivendra Mani Tripathi
Faculty of Pharmaceutical Sciences
Mahayogi Gorakhnath University Gorakhpur
Uttar Pradesh
India

Sudhanshu Mishra
Faculty of Pharmaceutical Sciences
Mahayogi Gorakhnath University Gorakhpur
Uttar Pradesh
India

Rishabha Malviya
Department of Pharmacy
School of Medical and Allied Sciences
Galgotias University, Greater Noida
Uttar Pradesh, India

&

Smriti Ojha
Department of Pharmaceutical Science
& Technology, Madan Mohan Malaviya
University of Technology, Gorakhpur
Uttar Pradesh, India

CHAPTER 1

Overview of Bioactive Compounds for Neurodegenerative Disorders

Abstract: The term "neurodegenerative disorders" refers to a broad category of pathological ailments that occur from increasing damage to the connections between the nervous system and the neurons. These disorders primarily affect neuronal malfunction and cause issues with strength, mobility, cognition, coordination, and sensation. Because of the body's need for and use of oxygen, the brain is susceptible to a variety of challenges, most notably oxidative stress. Potential neuroprotective drugs for the treatment of neurodegenerative disorders have been identified in natural products. Natural products derived from plants, animals, and fungi as well as the bioactive compounds they contain have been thoroughly investigated and studied in recent years for potential therapeutic uses against a range of disorders, including cancer, diabetes, hypertension, cardiovascular disease, and neurodegenerative diseases. The body's need for and usage of oxygen can lead to oxidative stress, which is one of the stresses that can affect the brain. The brain's high content of unsaturated fatty acids makes it more vulnerable. Crucial goals include addressing oxidative damage and developing effective and safe therapies for neurodegenerative diseases. Research into bioactive chemicals is essential to this endeavor.

Keywords: Bioactive compound, Hypertension, Neurodegenerative disorder, Natural compound, Oxidative stress.

INTRODUCTION

The term "neurodegenerative disorders" refers to a broad category of pathological ailments that occur from increasing damage to the connections between the nervous system and the neurons. The body's need for and use of oxygen makes the brain susceptible to a variety of challenges, most notably oxidative stress [1, 2]. The brain's high content of unsaturated fatty acids makes it more vulnerable. Pathogenesis establishes the parameters of typical pathogenic processes, of which apoptosis, oxidative stress, excitotoxicity, and conformational changes in neuronal proteins are of utmost relevance [3]. NDDs have no known effective medical treatment, and the existing strategy for managing these conditions has only been able to partially eliminate their clinical symptoms [4]. There are lots of potential

Shivendra Mani Tripathi, Sudhanshu Mishra, Rishabha Malviya & Smriti Ojha

outcomes in the investigation for natural compounds having bioactivity, and possessing various therapeutic properties like anti-inflammatory, anti-oxidant, and anti-tumorigenic activities, despite some recent advancements in chemical drug development [5]. In reaction to a severe insult traumatic event or ischemic brain damage, nerve cells are injured and mostly die, a condition known as acute neurodegeneration. The strong antioxidant and anti-inflammatory properties of naturally occurring bioactive substances make them interesting since they have many benefits for brain health [6].

Since long, natural products have been used and recognized for their therapeutic benefits. Phytoconstituents have been the subject of extensive research and exploration in recent years [7]. Numerous research works conducted in recent decades have documented the preventive properties of bioactive components against a range of disorders, including cancer, diabetes, heart disease, and neurological disorders. Potential neuroprotective drugs for the treatment of neurodegenerative disorders have been identified and evaluated for their therapeutic benefits in natural products [8].

Neurodegenerative Diseases

A variety of disorders known as neurodegenerative diseases (Fig. **1**) (NDD) are distinguished by continuing loss and selective malfunctioning of neurons and neural networks in the brain and spinal cord [9]. As a result, they can lead to a variety of issues, such as those involving mobility (known as ataxias), mental function (known as dementias), and the capacity to breathe, speak, and move. The aging of the world's population contributes to the rise in the prevalence of NDDs, which are crippling and incurable disorders [10]. Because the brain regulates several bodily functions, neurodegenerative diseases impact a variety of aspects of human functioning and impair the capacity to carry out both simple and complex tasks [11]. While in some cases, treatments aim to ameliorate symptoms, relieve pain if it is present, and/or restore movement and balance, most NDs develop without remission [12]. Neurodegenerative disorders are diverse ailments that cause neurons, glial cells, and their networks in the human brain and spinal cord to selectively malfunction and gradually lose their function [13]. Progressive loss of specific vulnerable neuronal populations is a hallmark of neurodegenerative disorders; this is in contrast to select static neuronal loss resulting from metabolic or toxic disorders [14]. Research has identified several similar processes *via* which dementia occurs, including the buildup of insoluble protein aggregates, apoptosis, necrosis, excitotoxicity, and neuroinflammation, despite the wide range of causes of neurodegenerative disorders [15]. Neurodegeneration is also influenced by downstream oxidative stress, decreased autophagy/lysosomal activity, and dysfunctional mitochondria [16, 17].

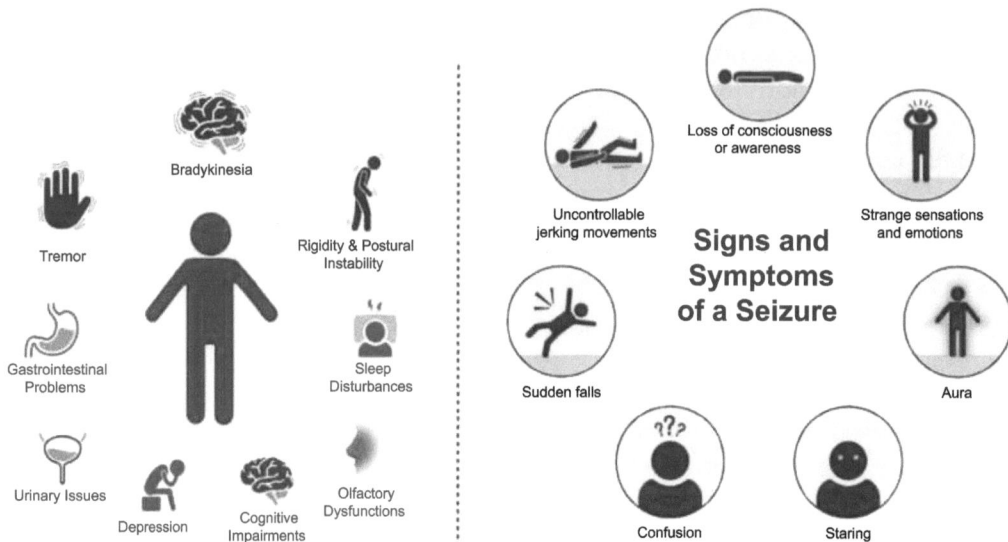

Fig. (1). Signs and symptoms of some neurological disorders.

Alzheimer's Disease (AD) and other Dementias

It is believed that the aberrant accumulation of proteins in and around brain cells is what causes Alzheimer's disease. Neurofibrillary tangles (NFTs), which are made of hyperphosphorylated tau protein, and amyloid-β plaques are the main causes of Alzheimer's disease (AD), a progressive neurodegenerative illness [18]. In addition to environmental and lifestyle variables including age and head trauma, genetic factors like mutations in the APP and PSEN genes and the APOE ε4 allele also play a part in the development of AD [19]. While tau tangles hinder neuronal transport and cause cell death, amyloid-β buildup impairs synaptic function and causes neuroinflammation. Cognitive decline is the result of neuronal loss, particularly in regions such as the hippocampus [20]. Other characteristics include cerebral amyloid angiopathy, astrogliosis, neuropil threads, and dystrophic neurites. These conditions work together to cause vascular dysfunction, brain shrinkage, and the gradual loss of memory, behavior, and cognitive functions in AD [21]. Furthermore, AD is associated with neuropil threads, dystrophic neurites, associated astrogliosis, microglial activation, and cerebral amyloid angiopathy [22].

Parkinson's Disease

The loss of dopaminergic neurons in the substantia nigra, a part of the brain involved in motor coordination, is the main cause of Parkinson's disease (PD), a progressive neurodegenerative illness [23]. Although the precise etiology is still unknown, environmental variables including age, pesticide exposure, and head trauma are thought to have a role, as are genetic abnormalities (*e.g.*, LRRK2, SNCA, PARK7). The development of illness is also significantly influenced by neuroinflammation, oxidative stress, and mitochondrial dysfunction [24]. Bradykinesia (slowness of movement), rigidity (stiffness of muscles), postural instability (balance issues), and tremors (resting tremors) are all examples of motor symptoms of Parkinson's disease. Usually minor at first, these symptoms get worse with time and interfere with day-to-day activities [25]. Cognitive decline, mood disorders (such as anxiety and depression), and autonomic dysfunction (such as constipation and sleep difficulties) are examples of non-motor symptoms that may manifest concurrently with or even ahead of motor symptoms [26]. As the condition worsens, facial expressions may become less lively and speech may become slurred or soft—a characteristic known as a "masked face." [27]. Parkinson's disease (PD) is primarily caused by the gradual degradation of dopaminergic neurons, which interferes with dopamine transmission to the basal ganglia, which are in charge of coordinating voluntary movements [28]. Abnormal motor control and the distinctive motor symptoms of Parkinson's disease are caused by this dopamine shortage. Dopaminergic neuron death is caused by several mechanisms, including oxidative stress, mitochondrial failure, neuroinflammation, and protein misfolding (*e.g.*, buildup of alpha-synuclein). Within neurons, aberrant protein clumps called Lewy bodies develop, which disrupt biological processes and accelerate the death of neurons. As a result of these interrelated disease processes, symptoms gradually worsen and cause substantial impairment.

Prion Disease

Transmissible Spongiform Encephalopathies (TSEs), another name for prion diseases, are progressive, lethal neurodegenerative conditions brought on by the aberrant misfolding of cellular prion protein (PrPC) into PrPSc. One kind of protein called a prion can cause the brain's regular proteins to fold strangely. The aggregation of abnormally folded prion protein into massive amyloid plaques and fibrous structures linked to neurodegeneration is a basic molecular event in prion disorders. Toxic aggregates that impair brain function develop when these misfolded proteins build up and cause normal prion proteins to take on an aberrant shape. Cognitive decline, motor dysfunction, coordination issues, and behavioral abnormalities are among the symptoms, which get worse over time until death [29]. The buildup of prion aggregates, which results in neurodegeneration, spongiform alterations in the brain, and neuronal death, is the disease process. Prion disease can spread through contaminated, sporadic, or hereditary causes and are incurable.

Synucleinopathies

Alpha-synuclein protein abnormally accumulates in synucleinopathies, a class of neurodegenerative diseases that result in the creation of deadly intracellular aggregates known as Lewy bodies. These disorders, which include multiple system atrophy, Parkinson's disease, and dementia with Lewy bodies, are brought on by genetic abnormalities in the SNCA gene or other elements that impact the metabolism of alpha-synuclein. Symptoms vary by subtype but usually include autonomic abnormalities, cognitive impairment, and motor dysfunction (bradykinesia, stiffness, and tremors) [30, 31]. Pathogenetically, they result from disruptions in αSyn metabolism (insufficient degradation leading to oligomer formation, and excessive synthesis).

Misfolded alpha-synuclein accumulation is the primary cause of the disease mechanism. It interferes with cellular functions including protein breakdown and mitochondrial function, causes oxidative stress and neuroinflammation, and gradually damages neurons.

Motor Neuron Diseases

Muscle weakness, atrophy, and eventually paralysis are symptoms of Motor Neuron Disease (MND), a rare and progressive neurodegenerative illness brought on by the loss of motor neurons in the brain and spinal cord. It is caused by a mix of environmental and genetic variables; family instances are influenced by mutations in genes such as SOD1, C9orf72, and TARDBP. Muscle cramps, fasciculations, trouble swallowing and speaking, and respiratory weakness are some of the symptoms, which frequently get worse with time [32]. Protein misfolding, oxidative stress, excitotoxicity, neuroinflammation, and mitochondrial dysfunction all contribute to the degeneration of motor neurons, which in turn causes a breakdown in the nervous system-muscle connection. Treatments may help control symptoms and enhance quality of life even when the condition is incurable [33].

Huntington's Disease

A mutation in the HTT gene causes Huntington's disease (HD), a genetic neurodegenerative illness that results in the formation of a toxic polyglutamine (polyQ) tract in the huntingtin protein and an aberrant expansion of CAG trinucleotide repeats. HD, which is inherited in an autosomal dominant form, usually shows up between the ages of 30 and 50. It is characterized by mental problems like depression and psychosis, cognitive decline like dementia, and motor symptoms like chorea, dystonia, and bradykinesia [34]. Seizures, fast progression, and stiffness are early symptoms of juvenile HD. Oxidative stress and neuronal death result from the mutant huntingtin protein aggregating in neurons and interfering with transcription, mitochondrial activity, and cellular clearance systems. Progressive malfunction and severe impairment are the results of the degeneration, which starts in striatal neurons and progresses to other parts of the brain [35, 36].

Spinocerebellar Ataxia

A collection of progressive, hereditary neurological disorders known as Spinocerebellar Ataxia (SCA) are mostly brought on by genetic abnormalities. These mutations frequently involve expansions of the CAG trinucleotide repeats, which result in the formation of polyglutamine (polyQ) regions in proteins such as ataxin-1 and ataxin-3. These mutations impair protein function, which leads to neuronal death and cellular toxicity, especially in the cerebellum. Different genetic changes influencing different cellular pathways give birth to non-polyQ SCAs. Progressive loss of motor coordination, gait ataxia, decreased balance, dysarthria, and oculomotor abnormalities are the hallmarks of SCA. In its final stages, tremors, dysphagia, and cognitive deterioration are also observed [37, 38]. Protein aggregation, oxidative stress, mitochondrial dysfunction, and compromised autophagy are the driving forces behind the disease's specific degeneration of cerebellar Purkinje cells. Each subtype has different symptoms, which get worse with time and cause substantial impairment.

Spinal Muscular Atrophy

Mutations or deletions in the SMN1 (Survival Motor Neuron 1) gene on chromosome 5q result in inadequate synthesis of the SMN protein, which causes Spinal Muscular Atrophy (SMA), a hereditary neuromuscular condition. The maintenance of motor neurons, which carry messages from the spinal cord to muscles, depends on this protein. Progressive muscular weakness results from the degeneration of motor neurons caused by insufficient SMN protein. Due to the autosomal recessive nature of SMA, both parents must have faulty copies of the SMN1 gene [39, 40]. MA is categorized as categories 0 through 4 and varies in severity based on residual SMN protein levels. Muscle weakness, decreased tone (hypotonia), trouble breathing and swallowing, and delays in motor milestones like sitting and walking are some of the symptoms. While lesser types (such as Types 3 and 4) develop later and proceed more slowly, the most severe form, Type 1, appears in infancy and can cause respiratory failure. Long-term inactivity makes muscle atrophy worse. The absence of SMN protein leads to the degeneration of motor neurons in the spinal cord's anterior horn, which is the disease mechanism. This causes muscles to atrophy and become paralyzed due to axonal degeneration, programmed cell death, and a lack of neurological input.

The disease is exacerbated by other causes, such as poor RNA processing and mitochondrial malfunction [41].

Pathological Mechanisms Underlying Neurodegenerative Diseases

In Neurodegenerative Diseases (NDDs), neurons and their networks in the brain and spinal cord gradually malfunction and die, resulting in debilitating symptoms including cognitive decline, mobility problems, and compromised essential functions. The following are important pathways that underlie NDDs: oxidative stress, mitochondrial dysfunction, excitotoxicity, neuroinflammation, protein misfolding and aggregation, and defective autophagy [42] (Fig. **2**). In prion diseases, prion proteins are misfolded, whereas in Alzheimer's Disease (AD), tau tangles and amyloid-β plaques are involved. Huntington's disease and Spinal Muscular Atrophy (SMA) are caused by genetic mutations, while synucleinopathies, such as Parkinson's disease, are caused by aberrant alpha-synuclein aggregation [43]. Certain neuronal populations are the focus of spinocerebellar ataxias and motor neuron disorders, which affect motor and coordination abilities. The development of the illness is made worse by mitochondrial malfunction, persistent inflammation, and specific neural susceptibility; disturbed autophagy also leads to the buildup of harmful proteins. The majority of NDDs are chronic and incurable, despite the fact that certain treatments reduce symptoms [44]. This emphasizes the critical need for drugs that address these overlapping pathogenic processes.

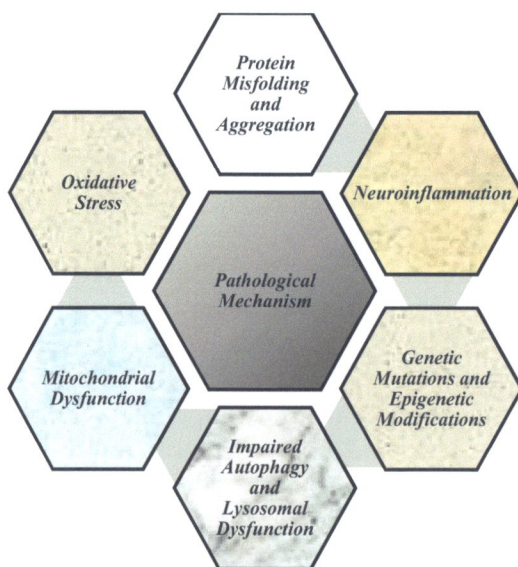

Fig. (2). Pathological mechanism of various neurological disorders.

Bioactive Compounds and their Role in NDD

The term "neurodegeneration" describes how nerve cells, or neurons, gradually deteriorate and lose their ability to function as well as their physical structure over time, ultimately leading to these cells' eventual demise. Several naturally occurring bioactive substances are medicinally used to treat neurodegenerative disorders (NDD) [45].

Phenols and Rolyphenols

The structural moieties of polyphenols, which are secondary plant metabolites, have one or more aromatic rings bearing hydroxyl (−OH) groups. Secondary plant metabolites known as polyphenols have a wide range of chemical configurations [46]. Among plant compounds, they make up the largest class. Connecting at least one aromatic ring to one or more hydroxyl functional groups is the fundamental structure of polyphenols. A good blood level of polyphenolic substances lowers the risk of cancer, NDD, and heart disease, as reflected by several research studies. By preventing microglial activation and the production of pro-inflammatory cytokines, polyphenols prevent neuroinflammation and shield neurons from damage brought on by inflammation. Polyphenols can lessen neuronal dysfunction by interfering with the aggregation of misfolded proteins, such as alpha-synuclein in Parkinson's disease and tau and amyloid-β in Alzheimer's disease. Polyphenols also improve synaptic plasticity, stimulate neurogenesis, maintain mitochondrial integrity, and control important neuroprotective signaling pathways including Nrf2 and sirtuins. Together, these benefits may protect against neurodegenerative processes, maintain neuronal function, and reduce cognitive decline [47].

Curcumin

Owing to a significant preclinical study, curcumin—a polyphenolic antioxidant produced from the turmeric root—has demonstrated remarkable efficiency in the treatment of inflammatory disorders, cancer, and wound repair. Dietary polyphenol curcumin, which is present in the curry spice turmeric, has strong anti-inflammatory and antioxidant qualities as well as the capacity to influence several targets that are connected to the etiology of chronic disorders [48]. It scavenges free radicals, protects neurons from oxidative stress, and suppresses inflammatory pathways, lowering neuroinflammation. Also, curcumin regulates protein aggregation, preventing alpha-synuclein aggregation in Parkinson's disease and encouraging the removal of amyloid-β plaques in Alzheimer's disease [49]. It also stimulates neurogenesis and synaptic plasticity, which enhance cognitive function, and it supports mitochondrial activity, reducing neuronal death. Despite its potential, curcumin's therapeutic efficacy is limited by its poor bioavailability and difficulty crossing the blood-brain barrier. Curcumin's inherent fluorescence makes it a valuable tool for tracking the course of disease, especially when examining retinal plaques, which may reflect the deposition of brain plaques [50].

Carotenoids

Carotenoids are colored substances that are made by plants, algae, and fungi but not by vertebrates. The primary carotenoids include α- and β-carotenes, β-cryptoxanthin, lycopene, zeaxanthin, and lutein [51]. With a linear hydrocarbon backbone (C40), these symmetrical tetraterpenes play significant roles in the organisms that create them. Carotenoids are biological in nature because of their unique structure and altering ability with biological cells or tissues especially towards reducing ROS. Carotenoids have well-established preventive functions in the human body, comprising the prevention and treatment of several diseases, including NDDs [52]. Carotenoids are naturally occurring pigments that are essential in preventing several brain disorders. These organic pigments have a significant role in the prevention of NDDs. There are known to be more than 600 carotenoids in nature, and they can be found in a vast array of organisms. The red, yellow, and orange pigments seen in a variety of animals and plants are caused by various kinds of carotenoids [53]. Due to their distinct structure, these compounds have a variety of bioactive properties, including antioxidant and anti-inflammatory properties. Through the prevention of neuro-inflammation, microglial activation, the excitotoxic pathway, autophagy modulation, attenuation of oxidative damage, and the activation of defensive antioxidant enzymes, these carotenoids offer neuroprotection. Histopathological analyses supported the significant improvements in body weight loss, neurobehavioral alterations decreased OS, and mitochondrial complex enzyme activity observed in the rat brain following oral lutein (an antioxidant) treatment. According to this study, lutein has the potential to be an effective treatment for NDD [54, 55].

Essential Oils

Essential oils comprise aromatic chemicals such as thymol, carvacrol, eugenol, cinnamic acid, limonene, and cinnamaldehyde as well as a few small-molecula--weight terpenoids. Their antibacterial capabilities are well-known [56]. Essential oils (EOs), together with their secondary metabolites and phenolic compounds, are among the most important components synthesized by aromatic plants. Essential oils that encourage neuroplasticity, improve memory and affect neurotransmitter systems include peppermint, lavender, and rosemary. They have antidepressant and anxiolytic properties that elevate mood and reduce anxiety. Essential oils can assist people with neurological disorders to manage their symptoms, improve their cognitive function, and feel better overall when used in aromatherapy or other applications [57].

Phytosterols

Phytosterols are a crucial dietary component. They are widely used in cosmetics and food products. As of now, we only know that they can decrease cholesterol, protect against cardiovascular disorders, and maybe fight cancer and inflammation [58]. It is important to do an in-depth study on the application of phytosterols for the therapeutic and preventive treatment of various disorders. Research into phytosterols' effects on the central nervous system is also quite fascinating. Phytosterols are the most prevalent plant sterols, with a structure that is similar to that of cholesterol. Sterols include brassicasterol, ergosterol, sitosterol, stigmasterol, campesterol, and lupeol [59]. Phytosterols available in circulation can penetrate the BBB at significant levels and assemble irreversibly in the brain. Phytosterols protect neurons by stabilizing membranes, reducing excitotoxicity, and supporting mitochondrial function. They also lower blood cholesterol levels, potentially reducing the formation of harmful amyloid plaques in Alzheimer's [60]. Actually, *in vivo*, phytosterols dramatically decreased brain β amyloid peptide levels as well as β- and γ-secretase activities, indicating that a diet high in plant sterols could be helpful for neurodegenerative disorders [61].

Bioactive Compounds for Management of NDD: Combinatorial Approaches

Researchers have recently looked into combination regimens, which involve taking regular medications for neurodegenerative disorders and combining them with bioactive substances derived from plants to gain greater efficacy and avoid side effects. Nowadays, various neurodegenerative disorders and related issues are treated with prescriptions for medications [62]. However, the majority of these medications have moderate to severe adverse effects. Research on multi-vitamin complexes revealed that an increase in multi-vitamin complex blood level decreases ROS formation in AD patient cells while also improving their ability to eliminate ROS [63]. The results of the *in vivo* investigation showed that animal groups treated with donepezil and resveratrol combos had an improved antioxidant state, which was caused by increased Super-oxide-dismutase activity, suggesting that both medications function in concert. Table **1** represents various combination therapies and their pharmacological actions.

Table 1. Combination therapy for the management of NDD.

S. No.	Combination Therapy	Activity	References
1	Donepezil and Extra virgin olive oil	Regulates synaptic proteins and reduces neuroinflammation, significantly reduced amyloid β aggregation, and enhancing the benefits of donepezil.	[64]
2	Donepezil and Ginkgo ketoester	Exerts anti amnesic effects *via* antioxidation and concentration-dependent inhibition of acetylcholinesterase and butyrylcholinesterase.	[65]
3	*Ginkgo biloba* extract and donepezil	Treatment has fewer negative effects and is better than donepezil monotherapy.	[66]
4	Caffeic acid and donepezil	Combination prevented the lipid peroxidation caused by sodium nitroprusside.	[67]
5	Donepezil and resveratrol	An improved antioxidant status implies that both drugs work synergistically.	[68]
6	Donepezil and gallic acid	Rectified rising MDA levels and brain activities.	[69]
7	A mixture of *Santalum album*, *Illicium verum*, and *Polygala tenuifolia*, and donepezil	Improved memory and depression in a mouse model of AD.	[70]
8	Fangchinoline and huperzine A	Suppressed AChE more than the treatments given separately.	[71]
9	Harmine and donepezil	AChE inhibition in the cerebral cortex of mice.	[72]
10	Memantine and huperzine A	Superior AChE inhibitor	[73]
11	Huperzine A, *Convolvulus pluricaulis* and *Celastrus paniculatus*	Synergistic AChE inhibition	[74]

CONCLUSION

Natural products and their bioactive components have been the subject of extensive research and investigation. These attributes can improve mental well-being. Complicated brain disorders are marked by a variety of malfunctioning pathways. As the world's population ages and becomes older, it is expected that NDD will become more common. This is expected to be the first sign of dementia, the most frequent age-related brain condition. It is possible to diagnose memory and cognitive deficits during the preclinical phase and delay the onset of NDD by implementing appropriate measures such as cognitive training, physical exercise, and nutritional therapies. The main focus is evaluating how these bioactive substances can support the brain's restricted capacity for repair and regeneration, promote neurogenesis, provide neuroprotection, and lessen NDDs. The antioxidant activity of the potential phytoconstituents successfully decreased oxidative species in patients by increasing antioxidant capacities at the cellular level, decreasing ROS production, and managing cytokine-induced inflammation. Our knowledge of the preventive benefits of phytoconstituents will advance as we get a deeper comprehension of the molecular/cellular level of action of these chemicals and their roles in neuron protection and in lowering oxidative stress. Bioactive substances may play a role in controlling gene expression in the brain and neuronal signaling pathways. The neuroprotective benefits of bioactive have been supported by published data thus far; however, further research is still required before we can officially introduce them as neuromedicines. Natural constituents are therefore preferred since they contain bioactive compounds with a variety of mechanisms of action that have neuroprotective effects.

REFERENCES

[1]   Dugger, B.N.; Dickson, D.W. Pathology of neurodegenerative diseases. *Cold Spring Harb. Perspect. Biol.,* **2017**, *9*(7), a028035.
[http://dx.doi.org/10.1101/cshperspect.a028035] [PMID: 28062563]

[2]   Olufunmilayo, E.O.; Gerke-Duncan, M.B.; Holsinger, R.M.D. Oxidative Stress and Antioxidants in Neurodegenerative Disorders. *Antioxidants,* **2023**, *12*(2), 517.
[http://dx.doi.org/10.3390/antiox12020517] [PMID: 36830075]

[3]   Jantas, D.; Lasoń, W. Preclinical evidence for the interplay between oxidative stress and rip1-dependent cell death in neurodegeneration: State of the art and possible therapeutic implications. *Antioxidants,* **2021**, *10*(10), 1518.
[http://dx.doi.org/10.3390/antiox10101518] [PMID: 34679652]

[4]   Ezike, T.C.; Okpala, U.S.; Onoja, U.L.; Nwike, C.P.; Ezeako, E.C.; Okpara, O.J.; Okoroafor, C.C.; Eze, S.C.; Kalu, O.L.; Odoh, E.C.; Nwadike, U.G.; Ogbodo, J.O.; Umeh, B.U.; Ossai, E.C.; Nwanguma, B.C. Advances in drug delivery systems, challenges and future directions. *Heliyon,* **2023**, *9*(6), e17488.
[http://dx.doi.org/10.1016/j.heliyon.2023.e17488] [PMID: 37416680]

[5]   Mucha, P.; Skoczyńska, A.; Małecka, M.; Hikisz, P.; Budzisz, E. Overview of the antioxidant and anti-inflammatory activities of selected plant compounds and their metal ions complexes. *Molecules,* **2021**, *26*(16), 4886.
[http://dx.doi.org/10.3390/molecules26164886] [PMID: 34443474]

[6]   Begdache, L.; Marhaba, R. Bioactive Compounds for Customized Brain Health: What Are We and Where Should We Be Heading? *Int. J. Environ. Res. Public Health,* **2023**, *20*(15), 6518.
[http://dx.doi.org/10.3390/ijerph20156518] [PMID: 37569058]

[7]   Chaachouay, N.; Zidane, L. Plant-Derived Natural Products: A Source for Drug Discovery and Development. *Drugs and Drug Candidates,* **2024**, *3*(1), 184-207.
[http://dx.doi.org/10.3390/ddc3010011]

[8]   Rahman, M.H.; Bajgai, J.; Fadriquela, A.; Sharma, S.; Trinh, T.T.; Akter, R.; Jeong, Y.J.; Goh, S.H.; Kim, C.S.; Lee, K.J. Therapeutic potential of natural products in treating neurodegenerative disorders and their future prospects and challenges. *Molecules,* **2021**, *26*(17), 5327.
[http://dx.doi.org/10.3390/molecules26175327] [PMID: 34500759]

[9]   Chakraborty, P.; Bhattacharyya, C.; Sahu, R.; Dua, T.K.; Kandimalla, R.; Dewanjee, S. Polymeric nanotherapeutics: An emerging therapeutic approach for the management of neurodegenerative disorders. *J. Drug Deliv. Sci. Technol.,* **2024**, *91*, 105267.
[http://dx.doi.org/10.1016/j.jddst.2023.105267]

[10]   Bilbul, M.; Finkel, J. *Major Neurocognitive Disorder*; Psychiatr. Consult, **2023**, pp. 219-239.
[http://dx.doi.org/10.1007/978-3-031-26837-3_13]

[11]   Azam, S.; Haque, M.E.; Balakrishnan, R.; Kim, I.S.; Choi, D.K. The Ageing Brain: Molecular and Cellular Basis of Neurodegeneration. *Front. Cell Dev. Biol.,* **2021**, *9*, 683459.
[http://dx.doi.org/10.3389/fcell.2021.683459] [PMID: 34485280]

[12]   Ramakrishnan, P.; Joshi, A.; Fazil, M.; Yadav, P. A comprehensive review on therapeutic potentials of photobiomodulation for neurodegenerative disorders. *Life Sci.,* **2024**, *336*, 122334.
[http://dx.doi.org/10.1016/j.lfs.2023.122334] [PMID: 38061535]

[13]   Yadav, D.K. Potential Therapeutic Strategies of Phytochemicals in Neurodegenerative Disorders. *Curr. Top. Med. Chem.,* **2021**, *21*(31), 2814-2838.
[http://dx.doi.org/10.2174/1568026621666211201150217] [PMID: 34852737]

[14]   Pandya, V.A.; Patani, R. Region-specific vulnerability in neurodegeneration: lessons from normal ageing. *Ageing Res. Rev.,* **2021**, *67*, 101311.
[http://dx.doi.org/10.1016/j.arr.2021.101311] [PMID: 33639280]

[15]   Wilson, D.M., III; Cookson, M.R.; Van Den Bosch, L.; Zetterberg, H.; Holtzman, D.M.; Dewachter, I. Hallmarks of neurodegenerative diseases. *Cell,* **2023**, *186*(4), 693-714.
[http://dx.doi.org/10.1016/j.cell.2022.12.032] [PMID: 36803602]

[16]   Bloomingdale, P.; Karelina, T.; Ramakrishnan, V.; Bakshi, S.; Véronneau-Veilleux, F.; Moye, M.; Sekiguchi, K.; Meno-Tetang, G.; Mohan, A.; Maithreye, R.; Thomas, V.A.; Gibbons, F.; Cabal, A.; Bouteiller, J.M.; Geerts, H. Hallmarks of neurodegenerative disease: A systems pharmacology perspective. *CPT Pharmacometrics Syst. Pharmacol.,* **2022**, *11*(11), 1399-1429.
[http://dx.doi.org/10.1002/psp4.12852] [PMID: 35894182]

[17]   Dong, X.; Wang, Y.; Qin, Z. Molecular mechanisms of excitotoxicity and their relevance to pathogenesis of neurodegenerative diseases. *Acta Pharmacol. Sin.,* **2009**, *30*(4), 379-387.
[http://dx.doi.org/10.1038/aps.2009.24] [PMID: 19343058]

[18]   Rajmohan, R.; Reddy, P.H. Amyloid-Beta and Phosphorylated Tau Accumulations Cause Abnormalities at Synapses of Alzheimer's disease Neurons. *J. Alzheimers Dis.,* **2017**, *57*(4), 975-999.
[http://dx.doi.org/10.3233/JAD-160612] [PMID: 27567878]

[19]   Zhang, H.; Li, X.; Wang, X.; Xu, J.; Elefant, F.; Wang, J. Cellular response to β-amyloid neurotoxicity in Alzheimer's disease and implications in new therapeutics. *Animal Model. Exp. Med.,* **2023**, *6*(1), 3-9.
[http://dx.doi.org/10.1002/ame2.12313] [PMID: 36872303]

[20]   Ma, C.; Hong, F.; Yang, S. Amyloidosis in Alzheimer's Disease: Pathogeny, Etiology, and Related Therapeutic Directions. *Molecules,* **2022**, *27*(4), 1210.
[http://dx.doi.org/10.3390/molecules27041210] [PMID: 35209007]

[21]   Gulisano, W; Maugeri, D; Baltrons, MA; Fà, M; Amato, A; Palmeri, A Erratum: role of amyloid-β and tau proteins in alzheimer's disease: confuting the amyloid cascade. *J. Alzheimer's Dis.,* **2018**, *68*, 415.
[http://dx.doi.org/10.3233/JAD-179935]

[22]   Long, J.M.; Holtzman, D.M. Alzheimer Disease: An Update on Pathobiology and Treatment Strategies. *Cell,* **2019**, *179*(2), 312-339.
[http://dx.doi.org/10.1016/j.cell.2019.09.001] [PMID: 31564456]

[23]   Zhou, Z.D.; Yi, L.X.; Wang, D.Q.; Lim, T.M.; Tan, E.K. Role of dopamine in the pathophysiology of Parkinson's disease. *Transl. Neurodegener.,* **2023**, *12*(1), 44.
[http://dx.doi.org/10.1186/s40035-023-00378-6] [PMID: 37718439]

[24]   Kim, T.W.; Koo, S.Y.; Studer, L. Pluripotent Stem Cell Therapies for Parkinson Disease: Present Challenges and Future Opportunities. *Front. Cell Dev. Biol.,* **2020**, *8*, 729.
[http://dx.doi.org/10.3389/fcell.2020.00729] [PMID: 32903681]

[25] Guerra, A.; Colella, D.; Giangrosso, M.; Cannavacciuolo, A.; Paparella, G.; Fabbrini, G.; Suppa, A.; Berardelli, A.; Bologna, M. Driving motor cortex oscillations modulates bradykinesia in Parkinson's disease. *Brain,* **2022**, *145*(1), 224-236.
[http://dx.doi.org/10.1093/brain/awab257] [PMID: 34245244]

[26] Carroll, V.; Rossiter, R.; Blanchard, D. Non-motor symptoms of Parkinson's disease. *Aust. J. Gen. Pract.,* **2021**, *50*(11), 812-817.
[http://dx.doi.org/10.31128/AJGP-07-21-6093] [PMID: 34713279]

[27] Aradi, S.D.; Hauser, R.A. Medical Management and Prevention of Motor Complications in Parkinson's Disease. *Neurotherapeutics,* **2020**, *17*(4), 1339-1365.
[http://dx.doi.org/10.1007/s13311-020-00889-4] [PMID: 32761324]

[28] Ramesh, S.; Arachchige, A.S.P.M. Depletion of dopamine in Parkinson's disease and relevant therapeutic options: A review of the literature. *AIMS Neurosci.,* **2023**, *10*(3), 200-231.
[http://dx.doi.org/10.3934/Neuroscience.2023017] [PMID: 37841347]

[29] Orge, L.; Lima, C.; Machado, C.; Tavares, P.; Mendonça, P.; Carvalho, P.; Silva, J.; Pinto, M.L.; Bastos, E.; Pereira, J.C.; Gonçalves-Anjo, N.; Gama, A.; Esteves, A.; Alves, A.; Matos, A.C.; Seixas, F.; Silva, F.; Pires, I.; Figueira, L.; Vieira-Pinto, M.; Sargo, R.; Pires, M.A. Neuropathology of animal prion diseases. *Biomolecules,* **2021**, *11*(3), 466.
[http://dx.doi.org/10.3390/biom11030466] [PMID: 33801117]

[30] Praschberger, R.; Kuenen, S.; Schoovaerts, N.; Kaempf, N.; Singh, J.; Janssens, J.; Swerts, J.; Nachman, E.; Calatayud, C.; Aerts, S.; Poovathingal, S.; Verstreken, P. Neuronal identity defines α-synuclein and tau toxicity. *Neuron,* **2023**, *111*(10), 1577-1590.e11.
[http://dx.doi.org/10.1016/j.neuron.2023.02.033] [PMID: 36948206]

[31] Valera, E.; Monzio Compagnoni, G.; Masliah, E. Review: Novel treatment strategies targeting alpha-synuclein in multiple system atrophy as a model of synucleinopathy. *Neuropathol. Appl. Neurobiol.,* **2016**, *42*(1), 95-106.
[http://dx.doi.org/10.1111/nan.12312] [PMID: 26924723]

[32] Foster, L.A.; Salajegheh, M.K. Motor Neuron Disease: Pathophysiology, Diagnosis, and Management. *Am. J. Med.,* **2019**, *132*(1), 32-37.
[http://dx.doi.org/10.1016/j.amjmed.2018.07.012] [PMID: 30075105]

[33] Oliver, D.J. Palliative care in motor neurone disease: where are we now? *Palliat. Care,* **2019**, *12*, 1178224218813914.
[http://dx.doi.org/10.1177/1178224218813914] [PMID: 30718958]

[34] McColgan, P.; Tabrizi, S.J. Huntington's disease: a clinical review. *Eur. J. Neurol.,* **2018**, *25*(1), 24-34.
[http://dx.doi.org/10.1111/ene.13413] [PMID: 28817209]

[35] Bakels, H.S.; Roos, R.A.C.; van Roon-Mom, W.M.C.; de Bot, S.T. Juvenile-Onset Huntington Disease Pathophysiology and Neurodevelopment: A Review. *Mov. Disord.,* **2022**, *37*(1), 16-24.
[http://dx.doi.org/10.1002/mds.28823] [PMID: 34636452]

[36] Jurcau, A. Molecular Pathophysiological Mechanisms in Huntington's Disease. *Biomedicines,* **2022**, *10*(6), 1432.
[http://dx.doi.org/10.3390/biomedicines10061432] [PMID: 35740453]

[37] Manto, M.; Serrao, M.; Filippo Castiglia, S.; Timmann, D.; Tzvi-Minker, E.; Pan, M.K.; Kuo, S.H.; Ugawa, Y. Neurophysiology of cerebellar ataxias and gait disorders. *Clin. Neurophysiol. Pract.,* **2023**, *8*, 143-160.
[http://dx.doi.org/10.1016/j.cnp.2023.07.002] [PMID: 37593693]

[38] Sullivan, R.; Yau, W.Y.; O'Connor, E.; Houlden, H. Spinocerebellar ataxia: an update. *J. Neurol.,* **2019**, *266*(2), 533-544.
[http://dx.doi.org/10.1007/s00415-018-9076-4] [PMID: 30284037]

[39]   Day, J.W.; Howell, K.; Place, A.; Long, K.; Rossello, J.; Kertesz, N.; Nomikos, G. Advances and limitations for the treatment of spinal muscular atrophy. *BMC Pediatr.,* **2022**, *22*(1), 632.
[http://dx.doi.org/10.1186/s12887-022-03671-x] [PMID: 36329412]

[40]   Bowerman, M.; Becker, C.G.; Yáñez-Muñoz, R.J.; Ning, K.; Wood, M.J.A.; Gillingwater, T.H.; Talbot, K. Therapeutic strategies for spinal muscular atrophy: SMN and beyond. *Dis. Model. Mech.,* **2017**, *10*(8), 943-954.
[http://dx.doi.org/10.1242/dmm.030148] [PMID: 28768735]

[41]   Rossoll, W.; Bassell, G.J. Spinal muscular atrophy and a model for survival of motor neuron protein function in axonal ribonucleoprotein complexes. *Results Probl. Cell Differ.,* **2009**, *48*, 87-107.
[http://dx.doi.org/10.1007/400_2009_4] [PMID: 19343312]

[42]   Clemente-Suárez, V.; Redondo-Flórez, L.; Beltrán-Velasco, A.; Ramos-Campo, D.; Belinchón-deMiguel, P.; Martinez-Guardado, I.; Dalamitros, A.; Yáñez-Sepúlveda, R.; Martín-Rodríguez, A.; Tornero-Aguilera, J. Mitochondria and Brain Disease: A Comprehensive Review of Pathological Mechanisms and Therapeutic Opportunities. *Biomedicines,* **2023**, *11*(9), 2488.
[http://dx.doi.org/10.3390/biomedicines11092488] [PMID: 37760929]

[43]   Sharma, S.; Young, R.J.; Chen, J.; Chen, X.; Oh, E.C.; Schiller, M.R. Minimotifs dysfunction is pervasive in neurodegenerative disorders. *Alzheimers Dement. (N. Y.),* **2018**, *4*(1), 414-432.
[http://dx.doi.org/10.1016/j.trci.2018.06.005] [PMID: 30225339]

[44]   Picca, A.; Calvani, R.; Coelho-Júnior, H.J.; Landi, F.; Bernabei, R.; Marzetti, E. Mitochondrial dysfunction, oxidative stress, and neuroinflammation: Intertwined roads to neurodegeneration. *Antioxidants,* **2020**, *9*(8), 647.
[http://dx.doi.org/10.3390/antiox9080647] [PMID: 32707949]

[45]   Mursal, M.; Kumar, A.; Hasan, S.M.; Hussain, S.; Singh, K.; Kushwaha, S.P.; Arif, M.; Kumar Singh, R.; Singh, D.; Mohammad, A.; Fatima, S.K. Role of natural bioactive compounds in the management of neurodegenerative disorders. *Intelligent Pharmacy,* **2024**, *2*(1), 102-113.
[http://dx.doi.org/10.1016/j.ipha.2023.09.006]

[46]   Rathod, N.B.; Elabed, N.; Punia, S.; Ozogul, F.; Kim, S.K.; Rocha, J.M. Recent Developments in Polyphenol Applications on Human Health: A Review with Current Knowledge. *Plants,* **2023**, *12*(6), 1217.
[http://dx.doi.org/10.3390/plants12061217] [PMID: 36986905]

[47]   Iqbal, I.; Wilairatana, P.; Saqib, F.; Nasir, B.; Wahid, M.; Latif, M.F.; Iqbal, A.; Naz, R.; Mubarak, M.S. Plant Polyphenols and Their Potential Benefits on Cardiovascular Health: A Review. *Molecules,* **2023**, *28*(17), 6403.
[http://dx.doi.org/10.3390/molecules28176403] [PMID: 37687232]

[48]   Sharifi-Rad, J.; Rayess, Y.E.; Rizk, A.A.; Sadaka, C.; Zgheib, R.; Zam, W.; Sestito, S.; Rapposelli, S.; Neffe-Skocińska, K.; Zielińska, D.; Salehi, B.; Setzer, W.N.; Dosoky, N.S.; Taheri, Y.; El Beyrouthy, M.; Martorell, M.; Ostrander, E.A.; Suleria, H.A.R.; Cho, W.C.; Maroyi, A.; Martins, N. Turmeric and Its Major Compound Curcumin on Health: Bioactive Effects and Safety Profiles for Food, Pharmaceutical, Biotechnological and Medicinal Applications. *Front. Pharmacol.,* **2020**, *11*, 01021.
[http://dx.doi.org/10.3389/fphar.2020.01021] [PMID: 33041781]

[49]   Pandey, N.; Strider, J.; Nolan, W.C.; Yan, S.X.; Galvin, J.E. Curcumin inhibits aggregation of α-synuclein. *Acta Neuropathol.,* **2008**, *115*(4), 479-489.
[http://dx.doi.org/10.1007/s00401-007-0332-4] [PMID: 18189141]

[50]   Manna, J.; Dunbar, G.L.; Maiti, P. Curcugreen treatment prevented splenomegaly and other peripheral organ abnormalities in 3xtg and 5xfad mouse models of alzheimer's disease. *Antioxidants,* **2021**, *10*(6), 899.
[http://dx.doi.org/10.3390/antiox10060899] [PMID: 34199639]

[51]   Milani, A.; Basirnejad, M.; Shahbazi, S.; Bolhassani, A. Carotenoids: biochemistry, pharmacology and treatment. *Br. J. Pharmacol.,* **2017**, *174*(11), 1290-1324.

[http://dx.doi.org/10.1111/bph.13625] [PMID: 27638711]

[52]   Gandla, K.; Babu, A.K.; Unnisa, A.; Sharma, I.; Singh, L.P.; Haque, M.A.; Dashputre, N.L.; Baig, S.; Siddiqui, F.A.; Khandaker, M.U.; Almujally, A.; Tamam, N.; Sulieman, A.; Khan, S.L.; Emran, T.B. Carotenoids: Role in Neurodegenerative Diseases Remediation. *Brain Sci.,* **2023**, *13*(3), 457.
[http://dx.doi.org/10.3390/brainsci13030457] [PMID: 36979267]

[53]   González-Peña, M.A.; Ortega-Regules, A.E.; Anaya de Parrodi, C.; Lozada-Ramírez, J.D. Chemistry, Occurrence, Properties, Applications, and Encapsulation of Carotenoids—A Review. *Plants,* **2023**, *12*(2), 313.
[http://dx.doi.org/10.3390/plants12020313] [PMID: 36679026]

[54]   Feng, J.; Zheng, Y.; Guo, M.; Ares, I.; Martínez, M.; Lopez-Torres, B.; Martínez-Larrañaga, M.R.; Wang, X.; Anadón, A.; Martínez, M.A. Oxidative stress, the blood–brain barrier and neurodegenerative diseases: The critical beneficial role of dietary antioxidants. *Acta Pharm. Sin. B,* **2023**, *13*(10), 3988-4024.
[http://dx.doi.org/10.1016/j.apsb.2023.07.010] [PMID: 37799389]

[55]   Babazadeh, A.; Vahed, F.M.; Liu, Q.; Siddiqui, S.A.; Kharazmi, M.S.; Jafari, S.M. Natural Bioactive Molecules as Neuromedicines for the Treatment/Prevention of Neurodegenerative Diseases. *ACS Omega,* **2023**, *8*(10), 3667-3683.
[http://dx.doi.org/10.1021/acsomega.2c06098] [PMID: 36743024]

[56]   Bahrami, A.; Delshadi, R.; Assadpour, E.; Jafari, S.M.; Williams, L. Antimicrobial-loaded nanocarriers for food packaging applications. *Adv. Colloid Interface Sci.,* **2020**, *278*, 102140.
[http://dx.doi.org/10.1016/j.cis.2020.102140] [PMID: 32171115]

[57]   Parham, S.; Kharazi, A.Z.; Bakhsheshi-Rad, H.R.; Nur, H.; Ismail, A.F.; Sharif, S.; RamaKrishna, S.; Berto, F. Antioxidant, antimicrobial and antiviral properties of herbal materials. *Antioxidants,* **2020**, *9*(12), 1309.
[http://dx.doi.org/10.3390/antiox9121309] [PMID: 33371338]

[58]   Sharma, N.; Tan, M.A.; An, S.S.A. Phytosterols: Potential metabolic modulators in neurodegenerative diseases. *Int. J. Mol. Sci.,* **2021**, *22*(22), 12255.
[http://dx.doi.org/10.3390/ijms222212255] [PMID: 34830148]

[59]   Bakrim, S.; Benkhaira, N.; Bourais, I.; Benali, T.; Lee, L.H.; El Omari, N.; Sheikh, R.A.; Goh, K.W.; Ming, L.C.; Bouyahya, A. Health Benefits and Pharmacological Properties of Stigmasterol. *Antioxidants,* **2022**, *11*(10), 1912.
[http://dx.doi.org/10.3390/antiox11101912] [PMID: 36290632]

[60]   Teibo John, ; Bello Samuel, ; Olagunju Abolaji, ; Olorunfemi Folashade, ; Ajao Oyetooke, ; Fabunmi Oluwatosin, Functional foods and bioactive compounds: Roles in the prevention, treatment and management of neurodegenerative diseases. *GSC Biol. Pharm. Sci.,* **2020**, *11*(2), 297-313.
[http://dx.doi.org/10.30574/gscbps.2020.11.2.0143]

[61]   Vanmierlo, T.; Rutten, K.; Dederen, J.; Bloks, V.W.; van Vark-van der Zee, L.C.; Kuipers, F.; Kiliaan, A.; Blokland, A.; Sijbrands, E.J.G.; Steinbusch, H.; Prickaerts, J.; Lütjohann, D.; Mulder, M. Liver X receptor activation restores memory in aged AD mice without reducing amyloid. *Neurobiol. Aging,* **2011**, *32*(7), 1262-1272.
[http://dx.doi.org/10.1016/j.neurobiolaging.2009.07.005] [PMID: 19674815]

[62]   Shoaib, S.; Ansari, M.A.; Fatease, A.A.; Safhi, A.Y.; Hani, U.; Jahan, R.; Alomary, M.N.; Ansari, M.N.; Ahmed, N.; Wahab, S.; Ahmad, W.; Yusuf, N.; Islam, N. Plant-Derived Bioactive Compounds in the Management of Neurodegenerative Disorders: Challenges, Future Directions and Molecular Mechanisms Involved in Neuroprotection. *Pharmaceutics,* **2023**, *15*(3), 749.
[http://dx.doi.org/10.3390/pharmaceutics15030749] [PMID: 36986610]

[63]   Rao, YL; Ganaraja, B; Marathe, A; Manjrekar, PA; Joy, T; Ullal, S Comparison of malondialdehyde levels and superoxide dismutase activity in resveratrol and resveratrol/donepezil combination treatment groups in Alzheimer's disease induced rat model. *Biotech,* **2021**, *3*, 11.

[http://dx.doi.org/10.1007/s13205-021-02879-5]

[64]   Batarseh, Y.S.; Kaddoumi, A. Oleocanthal-rich extra-virgin olive oil enhances donepezil effect by reducing amyloid-β load and related toxicity in a mouse model of Alzheimer's disease. *J. Nutr. Biochem.*, **2018**, *55*, 113-123.
[http://dx.doi.org/10.1016/j.jnutbio.2017.12.006] [PMID: 29413486]

[65]   Zhang, J.; Wang, J.; Zhou, G.S.; Tan, Y.J.; Tao, H.J.; Chen, J.Q.; Pu, Z.J.; Ma, J.Y.; She, W.; Kang, A.; Zhu, Y.; Liu, P.; Zhu, Z.H.; Shi, X.Q.; Tang, Y.P.; Duan, J.A. Studies of the Anti-amnesic Effects and Mechanisms of Single and Combined Use of Donepezil and Ginkgo Ketoester Tablet on Scopolamine-Induced Memory Impairment in Mice. *Oxid. Med. Cell. Longev.*, **2019**, *2019*, 1-16.
[http://dx.doi.org/10.1155/2019/8636835] [PMID: 30911351]

[66]   Yancheva, S.; Ihl, R.; Nikolova, G.; Panayotov, P.; Schlaefke, S.; Hoerr, R. *Ginkgo biloba* extract EGb 761®, donepezil or both combined in the treatment of Alzheimer's disease with neuropsychiatric features: A randomised, double-blind, exploratory trial. *Aging Ment. Health,* **2009**, *13*(2), 183-190.
[http://dx.doi.org/10.1080/13607860902749057] [PMID: 19347685]

[67]   Agunloye, O.M.; Oboh, G. Modulatory effect of caffeic acid on cholinesterases inhibitory properties of donepezil. *J. Complement. Integr. Med.*, **2018**, *15*(1), 20170016.
[http://dx.doi.org/10.1515/jcim-2017-0016] [PMID: 28941354]

[68]   Rao, YL; Ganaraja, B; Suresh, PK; Joy, T; Ullal, SD; Manjrekar, PA Effect of resveratrol and combination of resveratrol and donepezil on the expression of microglial cells and astrocytes in Wistar albino rats of colchicine-induced Alzheimer's disease. *Biotech,* **2023**, *3*, 13.
[http://dx.doi.org/10.1007/s13205-023-03743-4]

[69]   Obafemi, T.O.; Owolabi, O.V.; Omiyale, B.O.; Afolabi, B.A.; Ojo, O.A.; Onasanya, A.; Adu, I.A.I.; Rotimi, D. Combination of donepezil and gallic acid improves antioxidant status and cholinesterases activity in aluminum chloride-induced neurotoxicity in Wistar rats. *Metab. Brain Dis.*, **2021**, *36*(8), 2511-2519.
[http://dx.doi.org/10.1007/s11011-021-00749-w] [PMID: 33978901]

[70]   Liu, Q.F.; Choi, H.; Son, T.; Kim, Y.M.; Kanmani, S.; Chin, Y.W.; Kim, S.N.; Kim, K.K.; Kim, K.W.; Koo, B.S. Co-Treatment with the Herbal Medicine SIP3 and Donepezil Improves Memory and Depression in the Mouse Model of Alzheimer's Disease. *Curr. Alzheimer Res.*, **2022**, *19*(3), 246-263.
[http://dx.doi.org/10.2174/1567205019666220413082130] [PMID: 35422218]

[71]   Kong, X.P.; Ren, H.Q.; Liu, E.Y.L.; Leung, K.W.; Guo, S.C.; Duan, R.; Dong, T.T.X.; Tsim, K.W.K. The Cholinesterase Inhibitory Properties of Stephaniae Tetrandrae Radix. *Molecules,* **2020**, *25*(24), 5914.
[http://dx.doi.org/10.3390/molecules25245914] [PMID: 33327436]

[72]   He, D.; Wu, H.; Wei, Y.; Liu, W.; Huang, F.; Shi, H.; Zhang, B.; Wu, X.; Wang, C. Effects of harmine, an acetylcholinesterase inhibitor, on spatial learning and memory of APP/PS1 transgenic mice and scopolamine-induced memory impairment mice. *Eur. J. Pharmacol.*, **2015**, *768*, 96-107.
[http://dx.doi.org/10.1016/j.ejphar.2015.10.037] [PMID: 26526348]

[73]   Shao, Z-Q. Comparison of the efficacy of four cholinesterase inhibitors in combination with memantine for the treatment of Alzheimer's disease. *Int. J. Clin. Exp. Med.*, **2015**, *8*(2), 2944-2948.
[PMID: 25932260]

[74]   Ahmad, I.; Swaroop, A.; Bagchi, D. A synergistic combination of Huperzine A, *Convolvulus pluricaulis* and *Celastrus paniculatus* promote cognitive function and health. *FASEB J.,* **2018**, *32*(S1)
[http://dx.doi.org/10.1096/fasebj.2018.32.1_supplement.656.7]

CHAPTER 2

Medicinal Plants and Their Phytoconstituents in Neurodegenerative Disorders

Abstract: The human brain is made up of neurons and neuroglia, the latter of which are essential for maintaining neurons and responding to injury. Brain activity impairment can result from malfunctioning neuroglia. Chronic neuroglial activation affects neurodegeneration, which has a major impact on brain aging and neuropathological diseases. There are intriguing neuroprotective chemicals in traditional medicine that may offer an effective option than synthetic drugs. Researchers can improve natural treatments for neurological disorders by isolating active phytochemicals while discovering more about their processes, doses, and safety profiles. A variety of food sources including marine algae have neuroprotective properties. Bioactive substances with anti-inflammatory and antioxidant properties are found in marine algae. Nuts, fruits, vegetables, and other foods include phytochemicals that can prevent neurodegeneration. Resveratrol, curcumin, sulforaphane, flavonoids, and organosulfur compounds are important substances that shield neurons by a variety of mechanisms, such as anti-inflammatory and antioxidant ones. In addition to supporting general brain health, a varied diet high in these neuroprotective chemicals may lower the risk of neurological disorders.

Keywords: Brain aging, Flavonoids, Human brain, Phytochemicals, Resveratrol.

INTRODUCTION

Neurons and neuroglia make up the extraordinarily complex structure of the human brain, with neurons being the primary source of nerve signal transmission [1]. Astrocytes and microglia are crucial for maintaining neurons and stepping in when they are harmed or under stress [2]. These neuroglia serve as watchdogs over the health of the neurons, and when they malfunction, it can seriously impair brain activity. Neuronal signals play a major role in neuroglial activation [3]. Acute injuries trigger responses that either aid in regenerating injured neurons or kill them if they are not viable; these responses are regarded as normal and neuroprotective. Chronic processes, however, have the potential to cause persistent neuroglial activation, which would impair their capacity to maintain homeostasis and possibly harm healthy neurons [4].

Shivendra Mani Tripathi, Sudhanshu Mishra, Rishabha Malviya & Smriti Ojha

An essential component of neuropathological disorders and brain aging is neurodegeneration. Brain pathology, which includes neurodegenerative and cerebrovascular illnesses, is a leading cause of death and cognitive impairment worldwide. Many neuropsychiatric and neurodegenerative conditions can be extremely crippling, including Parkinson's disease, depression, and Alzheimer's disease [5]. An important field of study is neuroprotection, which attempts to protect the central nervous system from both acute injuries (such as stroke or trauma) and chronic neurodegenerative diseases. Due to the scarcity of available treatments, stroke and dementia present substantial obstacles. This has prompted research into the mechanisms of neuronal death and the development of novel therapeutic agents. In the end, strengthening neuroprotection techniques can lessen the effects of neurological conditions and enhance general brain health [6].

Medicinal plants include natural phytochemicals that are frequently thought to be less hazardous than synthetic medications. However, because traditional herbal remedies were traditionally made from raw materials, there are questions about the precise therapeutic effects, repeatability, modes of action, and identities of the active ingredients [7]. The focus of recent research has switched from researching these plants as a whole to separating and analyzing their constituent parts. It is essential to discover and characterize the active compounds found in medicinal plants for treating neurological illnesses. This methodology enables scientists to evaluate their possible therapeutic benefits, particularly in the treatment of neurodegenerative illnesses [8]. Scientists hope to gain a better understanding of the mechanisms of action, dosage requirements, and safety profiles of particular phytoconstituents by isolating and researching them (Table **1**). This targeted approach may lead to the development of more effective and reliable natural remedies for neurological disorders while minimizing potential side effects associated with whole-herb preparations [9].

Table 1. Various phytoconstituents and their pharmacological activities.

Sr. No.	Active Constituents	Activity	Refs.
1.	Ascorbic acid	Anti-oxidant	[79]
2.	Acacetin/flavanoid	Antiparkinson	[80]
3.	Apigenin	Alzheimer	[81]
4.	Quercetin	Alzheimer/Parkinson	[82]
5.	Resveratrol	Alzheimer	[83]
6.	Sulforaphane	Anti-inflammatory	[84]
7.	Silibinin	Ischemia	[85]

Neurodegenerative Diseases

Progressive conditions of the central nervous system (CNS) that also impact the peripheral nervous system are known as neurodegenerative diseases. Neurodegeneration is the result of a slow and cumulative loss of brain cells, which characterizes them. Free radicals are the main causes of this process, and Reactive Oxygen Species (ROS) and Reactive Nitrogen Species (RNS) contribute significantly to it by producing them [10]. The development of different neurological disorders is also significantly influenced by neuroinflammatory processes. Over 600 neurological illnesses have been reported globally, according to the National Institute of Neurological Diseases and Stroke [11]. Anxiety encompasses mental, physical, emotional, and behavioral aspects of psychological and physiological states. It can interfere significantly with daily life, leading to apprehension about everyday events. Anxiety disorders include seven clinical conditions: Generalized Anxiety Disorder (GAD), Panic Disorder, Phobias, Agoraphobia, Social Anxiety Disorder (SAD), Obsessive-Compulsive Disorder (OCD), Post-Traumatic Stress Disorder (PTSD), and Separation Anxiety Disorder. Neurotransmitters such as dopamine, noradrenaline, serotonin, neuropeptides, neurosteroids, and cytokines play modulatory roles in anxiety states [12].

The neuropsychiatric disorder known as attention-deficit/hyperactivity disorder (ADHD) is characterized by symptoms like restlessness, mood fluctuations, disorganization, and difficulty focusing. It usually starts in childhood and lasts throughout adulthood [13]. Depression is characterized by a lowered mood and widespread unhappiness that can range from mild melancholy to complete agony. It is a prevalent emotional condition impacted by both environmental and biological variables [14]. According to the World Health Organization (WHO), depression is a major cause of morbidity and mortality, with an estimated 450 million individuals suffering from mental or behavioral disorders [15]. Depression is associated with structural alterations in neurons, including abnormalities in 5-HT and its receptors, dysfunction of the HPA axis, and decreased volume of the frontal cortex and hippocampus [16].

Conditions such as Alzheimer's disease (AD) and Parkinson's disease (PD) are classified as dementia. Parkinson's disease (PD) is a severe motor disorder characterized by bradykinesia, tremor, stiffness, and impaired balance as a result of dopaminergic neurons in the substantia nigra being lost [17]. Memory loss, cognitive impairment, and psychiatric problems are characteristics of AD [18].

About 50 million people worldwide suffer from epilepsy, a condition marked by recurrent seizures brought on by an overabundance of cerebral neuron discharge.

Neuronal damage and cell death are caused by reduced sodium and calcium channel activity, GABA-mediated inhibition, and excessive glutamate-mediated neurotransmission [19]. Excessive glutamate stimulation can cause neuronal cells to become injured or die, a condition known as excitotoxicity. It happens in diseases such as amyotrophic lateral sclerosis, multiple sclerosis, stroke, traumatic brain injury, and spinal cord injury. Excitotoxicity involves over-activation of NMDA receptors, leading to Ca^{2+} influx and excessive NO generation [20]. Schizophrenia is a severe psychiatric illness with positive symptoms (delusions, hallucinations, thought disorders) and negative symptoms (social withdrawal, emotional flattening). Neurotransmitters such as dopamine, 5-HT, acetylcholine, and norepinephrine levels are upregulated in this condition [21].

Endogenous neuroprotection is a component of neuroregeneration, which promotes neurite outgrowth and neuroplasticity. Restoring synaptic and neural networks is crucial for the recovery of brain function. According to recent research, stimulatory chemicals including Brain-Derived Neurotrophic Factor (BDNF) and Nerve Growth Factor (NGF) can help injured neurons repair. Neuroprotection entails preventing damage from neurotoxic species, reducing neuro-inflammatory responses, and enhancing cognitive performance [22]. The potential for phytochemicals derived from traditional herbal extracts to prevent neurodegeneration and modulate neuronal function suggests that these compounds may have positive effects on the vascular system and cerebrovascular blood flow [23].

Neuroprotective Mechanism of Medicinal Plant

The vital systems known as neuroprotective mechanisms are what keep neurons safe from deterioration, harm, and death. These processes may be extrinsic—triggered by outside stimuli such as drugs—or intrinsic, occurring naturally within the body. To restore neuronal function following injury, cell repair and regeneration mechanisms like neurogenesis and axonal regeneration are essential [24]. Neurons are shielded from oxidative stress and reactive oxygen species by defense mechanisms that are aided by antioxidant enzymes. Anti-inflammatory reactions control the immune system, lowering the central nervous system inflammation linked to neurodegenerative diseases. Protecting the mitochondria from dysfunction maintains their health, lowering oxidative stress and continuing the generation of energy. Neuroprotective systems control neurotransmitter levels and block excitotoxic pathways to prevent excitotoxicity [25]. Apoptosis inhibitors increase the survival of neurons under stressful circumstances by preventing excessive neuronal cell death. For treatments of neurodegenerative disorders, it is imperative to comprehend and utilize these pathways shown in Fig. (**1**), with natural chemicals produced from medicinal plants being one potential therapeutic avenue.

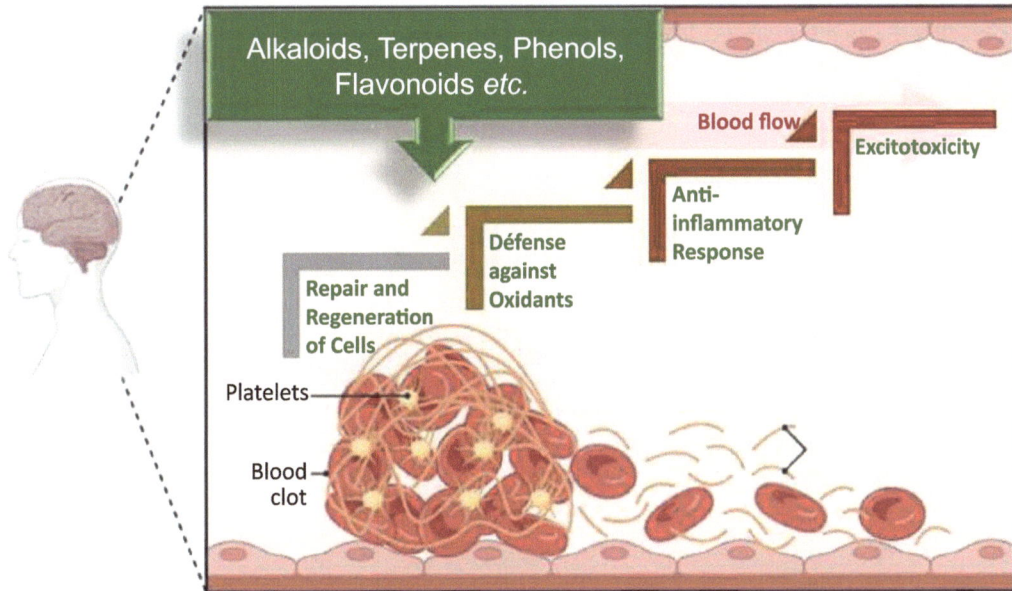

Fig. (1). Neuroprotective mechanisms of phytoconstituents.

Important mechanisms known as neuroprotective mechanisms guard neurons from deterioration, harm, and death. They include extrinsic, which is induced by outside sources such as drugs or treatment procedures, and intrinsic, which is innate to the organism. Essential components of neuroprotection include cell repair and regeneration, which include complex processes like neurogenesis—the production of new neurons from neural stem cells—and axonal regeneration—the regeneration of injured axons [26]. After injury or impairment, these pathways are essential for reestablishing neural connection and function. Furthermore, as neurons are especially susceptible to oxidative stress and damage from Reactive Oxygen Species (ROS), defense against oxidants is crucial. Antioxidant enzymes such as catalase and superoxide dismutase are examples of neuroprotective mechanisms that help neutralize Reactive Oxygen Species (ROS) and maintain redox balance, hence protecting neuronal integrity and function. Neuroprotective systems use an anti-inflammatory response to counteract oxidative stress and lessen immune-mediated neuronal damage [27]. Various neurodegenerative illnesses are directly linked to chronic inflammation in the central nervous system. Thus, the two most important tactics for maintaining neurological function are immune activity modulation and neural inflammation reduction. The production of anti-inflammatory cytokines and the suppression of pro-inflammatory mediators work together to reduce immune-mediated neuronal damage [28]. Furthermore, maintaining neuronal survival and averting neurodegeneration depend on defense against mitochondrial malfunction. A malfunctioning mitochondria can cause neural dysfunction and even death. Mitochondria are essential for energy production and cellular metabolism. By maintaining mitochondrial structure and function, improving energy output, and lowering oxidative stress inside mitochondria, neuroprotective processes work to promote mitochondrial health [29].

Excitotoxicity, a crucial component of neuroprotection, is the term used to describe the harm done to neurons when neurotransmitter receptors—especially glutamate—are overstimulated. By controlling neurotransmitter levels, boosting the action of inhibitory neurotransmitters like gamma-aminobutyric acid (GABA), and obstructing excitotoxic signaling pathways, neuroprotective mechanisms work against excitotoxicity. These systems preserve neuronal viability and guard against excitotoxic damage by limiting excessive neuronal excitement. Apoptosis inhibitors are also essential for reducing excessive neuronal cell death and increasing neuronal survival in stressful environments. Programmed cell death, or apoptosis, is a major factor in neurodegenerative diseases [30].

Echinacea purpurea

Purple coneflower, or *Echinacea purpurea* (L.) Moench (EP), is a perennial herbaceous blooming plant of the Asteraceae family. The United States is the birthplace of the Echinacea genus, however, its species can be found all over the world. Of the nine species of Echinacea, only three—*Echinacea angustifolia DC, Echinacea pallida Nutt.,* and *Echinacea purpurea* (L.) Moench—are used as medicinal herbs with a wide range of therapeutic purposes [31]. Numerous notable subgroups of bioactive compounds with pharmacological properties have been identified from the Echinacea species. The most important components of *Echinacea purpurea* (L.) Moench are alkylamides, polysaccharides, glycoproteins, flavonoids, and phenolic compounds, which include derivatives of caffeic acid like, chicoric acid, caftaric acid, chlorogenic acid, and echinacoside, and whose amounts vary depending on the sections of the plant. Researchers also discovered that phylloxanthobilins, -phellandrene, acetaldehyde, dimethyl sulphide, camphene, hexanal, -pinene, and limonene are present in all plant tissues, irrespective of species. The plant sections that are used determine the amount of terpenoids, aldehydes, and fatty acids present. Various pharmacologically significant advantages for human health, including neuroprotective and cardiovascular properties, have been found for echinacoside [32 - 34]. Research showed that rats' neurochemical and behavioral deficits caused by bifenthrin are prevented by *Echinacea purpurea* hydroethanolic extract

(EchEE). EchEE may be a neuroprotective agent against pesticide-induced brain dysfunction since it decreases oxidative stress, alters pro-inflammatory and apoptotic markers, and enhances neurotransmitter levels and behavioral performance [33].

Matricaria chamomilla L.

The Asteraceae family includes the well-known medicinal herb *Matricaria chamomilla*, which is most commonly known as chamomile. This annual plant can withstand the winter and grows well in a variety of soil conditions. *M. chamomilla's* native habitats are in northern and western Asia, as well as in southern and eastern Europe [35, 36]. According to various research works conducted in various regions, the most commonly used portion of *M. chamomilla* (Babonj/Babounj) in traditional Moroccan medicine is the flowers, followed by the leaves and the entire plant. For the treatment of diabetes, neurological diseases, diarrhea, angina, canker sores, abscesses, infections, and painful menstruation, it is made as an infusion or decoction. Studies on the phytochemical makeup of *M. chamomilla* extract and essential oil (EO) have identified over 120 components. Terpenoids made up the majority of the ingredients in M. chamomilla EO; the most important ones were -farnesene, chamazulene, bisabolol and its oxides A and B, and bisabolone oxide A [37, 38]. In rats suffering from cerebral ischemia/reperfusion damage, this study discovered that *Matricaria chamomilla* extract reduced motor dysfunction. It decreased blood levels of malondialdehyde (MDA), but it had no discernible impact on the brain's nitric oxide levels or antioxidant capacity [39].

St. John's Wort (Hypericum perforatum)

Worldwide, *Hypericum perforatum*, also known as St. John's wort, is a perennial plant that grows in many different places. It has long been utilized in conventional medicine to treat a wide range of illnesses, It is administered externally as an oil to heal wounds, burns, inflammation of the skin, and discomfort in the nerves. On the inside, it is acknowledged for how well it works to relieve mild to severe sadness and anxiety, frequently serving as a natural substitute for pharmaceutical antidepressant treatments. Hypericum perforatum has a wide variety of physiologically active chemicals [40, 41]. The quantities of these components are typically variable because of ecological factors, genetic variation within the species, adulteration, sample preparation and processing, storage conditions (such as light exposure and harvesting time), and adulteration. The most well-known natural products, hypericin and hyperforin, are derived from the Hypericum species. While glutamate and acetylcholine concentrations in the extracellular brain are raised and stress-induced behaviors are actively reduced by hypericin, hyperforin regulates the expression of genes associated with depression. Studies indicate that *Hypericum perforatum*, also known as St. John's Wort, and a few of its main constituents may offer neurotoxicity protection. In the setting of neurodegenerative diseases like Alzheimer's and Parkinson's disease, this putative preventive action might be advantageous [42, 43]. In a randomized, controlled trial, St. John's Wort extract (*Hypericum perforatum*) was found to be as effective as fluoxetine (Prozac) in treating mild to moderate depression, showing similar reductions in depression scores. Both treatments led to significant improvements with minor side effects, suggesting that Hypericum could be a viable alternative to synthetic antidepressants [44].

Boswellia Serrata (Salai Guggal)

The tree *Boswellia serrata* grows in India, Northern Africa, and the Middle East. Its resin, known as oleo-resin, contains active chemicals such as beta boswellic acid with various therapeutic uses. In addition to its analgesic properties, the resin exhibits anti-inflammatory, anti-arthritic, antirheumatic, antidiarrheal, antihyperlipidemic, antiasthmatic, anticancer, and antibacterial properties. According to studies, boswellic acid derivatives, including acetyl-keto-beta-boswellic acid (AKbetaBA), dramatically reduce the expression of NF-kappa--dependent genes related to inflammation [45]. This implies that there may be advantages to treating inflammatory chronic conditions like Alzheimer's disease (AD). As demonstrated by decreased amyloid plaques and changed biochemical indicators such as raised acetylcholine (ACh) and suppressed acetylcholine esterase (AChE), *B. serrata* boosted activity levels and improved cognitive performance in animal models of AD. Additionally, co-administration of ginger resulted in improved histopathologic and cognitive outcomes. *B. serrata* resin extract showed neuroprotective potential in Parkinson's disease (PD) by improving dopaminergic neuron survival and decreasing apoptosis [46, 47]. Studies on *B. serrata* administered by mothers revealed benefits in learning and memory in their offspring, which were associated with structural alterations in hippocampus neurons. These results indicate the possibility of using boswellia to improve cognitive abilities and lessen the effects of neurodegenerative diseases [48]. *Boswellia serrata* demonstrated a dose-dependent neuroprotective effect against 3-nitropropionic acid-induced Huntington's disease in rats, improving behavioral deficits, and memory, and reducing inflammatory biomarkers. The dose of 180 mg/kg showed the most significant protection, supported by histological findings, highlighting its potential as an adjuvant therapy for Huntington's disease [49].

Lepidium meyenii (Maca)

Maca, or *Lepidium meyenii* Walpers in scientific parlance, is a plant belonging to the Brassicaceae family that grows above 4,000 meters in the Peruvian Central Andes. Its pentane extract contains benzylisothiocyanates, macamides, macaenes, and lipo-soluble alkaloids. When injected intravenously, these substances can successfully reach the brain [50]. Studies have demonstrated that maca, at doses of 10 and 30 mg/kg, greatly lessens the neuronal damage caused by tumors in rats while maintaining normal dendritic cell morphologies. Additionally, a dose of 30 mg/kg lowers the risk of a heart attack and has neuroprotective effects. The aqueous fraction of Maca's methanolic extract is where the majority of its antioxidant activity is found. It has shown neurobiological benefits *in vitro*, improving cell survival and decreasing cytotoxicity against 6-OHDA-induced oxidative stress. Maca increases the activity of superoxide dismutase and scavenges free radicals such as nitric oxide, strengthening the balance of oxidation-reduction enzymes. Maca methanol extract increased the viability of PC12 cells exposed to neurotoxic damage from DA and 6-OHDA in a cell model of Parkinson's disease [51]. Pretreatment for 12 hours with 10 µg Maca extract greatly boosted cell viability. Maca may impede age-related cognitive impairment by increasing autophagy-related proteins and nitrogen-related pathways. An additional indication that Maca's neuroprotective benefits are related to lowering oxidative stress by stifling free radical activity is the rise in SOD activity. According to a study, giving middle-aged mice a 5-week Maca supplement improved their motor coordination, endurance, and cognitive performance. Increased autophagy-related proteins and improved mitochondrial respiratory efficiency probably caused this improvement. This shows that maca may be able to maintain brain health and stop cognitive decline [52]. Maca improved cognitive function, motor coordination, and endurance in middle-aged mice by enhancing mitochondrial activity and upregulating autophagy-related proteins in the cortex. These findings suggest maca's potential as a functional food to combat age-related cognitive decline [52].

Acorus calamus (Vacha)

Sweet flag (*Acorus calamus* Linn.) is a prominent medicinal plant in Ayurveda, used to cure neurological problems and other maladies. Vacha is the Sanskrit name for the plant. High amounts of the secondary metabolites α- and β-asarone, which have varied pharmacological properties such as antioxidant, neuroprotective, antidiabetic, anticancer, and antibacterial actions, are found in the rhizomes of *A. calamus* and related species [53]. *A. calamus* is grown all over the world in tropical and subtropical climates, and many different cultures have long valued its therapeutic properties. Numerous phytoconstituents, including alkaloids, tannins, glycosides, volatile oils, flavonoids, monoterpenes, steroids, sesquiterpenes, saponins, mucilage, and polyphenolic chemicals, are present in the plant. It is used to treat inflammatory conditions in Ayurveda and constipation and digestive problems in Chinese medicine [54, 55]. Research has demonstrated that *A. calamus* and its components can improve immune function and lessen immunosuppression brought on by stress. The plant contains monoterpenes such as limonene, α- and β-pinenes, myrcene, para-cymene, α-terpinene, β-phellandrene, gamma-terpinene, terpinolene, and thujane. The capacity of *A. calamus* to revitalize the brain, soothe the nervous system, and lessen anxiety and epileptic episodes is well recognized. Furthermore to a nutritious diet to prevent ailments like constipation, consuming 4g of vacha powder with honey every day is a suggested treatment for anxiety and epilepsy [56 - 58].

Vitis vinifera

The grapevine, *Vitis vinifera*, is often used to minimize the accumulation of tau tangles and amyloid plaques, which are linked to neurological disorders such as Alzheimer's. It increases cholinergic activities and reduces inflammation and oxidative damage [59, 60]. Extracts from *Vitis vinifera* are rich in natural substances called polyphenols, which have positive benefits on brain health strong anti-oxidative, anti-inflammatory, anti-acetylcholinesterase, and anti-amyloidogenic properties are exhibited by these polyphenols, which promote healthy aging of the brain. It reduces inflammation and oxidative stress, which helps shield neurons from harm. Its anti-amyloidogenic qualities stop the development of dangerous amyloid plaques, while its anti-acetylcholinesterase qualities promote cholinergic function, which is crucial for memory and learning. According to these results, *Vitis vinifera* extracts may be a useful natural strategy for maintaining cognitive function and averting neurodegenerative diseases as a part of normal aging [61 - 63]. Vitisin A, a resveratrol tetramer from *Vitis vinifera* stembark, improved cognitive and memory functions in scopolamine-induced models by restoring synaptic mechanisms and upregulating BDNF-CREB signaling [64].

Hyoscyamus niger

Henbane, or *Hyoscyamus niger*, is a plant that is high in tropane alkaloids like hyoscine and hyoscyamine but has very little L-DOPA. These substances are widely recognized for having strong anticholinergic properties [65, 66]. Parkinson's disease (PD) is characterized by impairments in motor function, including tremors, which are caused by a relative excess of acetylcholine due to a decrease in dopamine levels in the striatum. This excess acetylcholine is consequently countered and symptoms, especially tremors, can be managed using anticholinergic medications such as those produced by *H. niger*. These anticholinergic substances work to restore the proper balance of neurotransmitters in the brain, which helps PD patients live better lives and experience fewer motor symptoms. They do this by blocking the effect of acetylcholine [67]. The aqueous methanol extract of *Hyoscyamus niger* seeds, containing 0.03% L-DOPA, improved motor function and striatal dopamine levels in MPTP-induced Parkinson's disease models by inhibiting monoamine oxidase activity and reducing mitochondrial hydroxyl radical generation [68].

Celastrus paniculatus

The seeds and seed oil of *Celastrus paniculatus*, sometimes referred to as Jyotishmati or Malkangni in Ayurveda, are traditionally used as memory boosters and intelligence enhancers. Its benefits for brain function are mentioned in ancient Indian literature. The antioxidant activities of the aqueous seed extract may be responsible for some of its cognitive-enhancing qualities. Its inflorescences' methanol extract has anti-inflammatory properties that make it potentially useful in the treatment of neurodegenerative diseases (NDDs). Common NDDs include Huntington's disease (HD), Parkinson's disease (PD), and Alzheimer's disease (AD), which are caused by a complicated interaction between hereditary variables and aberrant cellular dynamics, including mitochondrial malfunction, oxidative stress, and unfolded protein accumulation [69, 70]. The majority of current therapies for these conditions are palliative in nature and do not stop the disease's progression, which emphasizes the need for novel therapeutic alternatives that can lessen neuronal damage and regulate the course of the disease. Plants have long been used to cure a wide range of illnesses, with *C. paniculatus* being particularly well-suited for treating neurological conditions. In Ayurveda, it has been used for millennia to treat a variety of conditions, including anxiety, insomnia, facial paralysis, dementia, epilepsy, and cognitive dysfunctions. The phytochemical properties, traditional use, and research on the possible therapeutic benefits of *C. paniculatus* seeds and seed oil against common neurological disorders are highlighted in this study [71, 72]. Aqueous extracts of *Celastrus paniculatus* seeds protected neuronal cells from glutamate-induced toxicity by attenuating mitochondrial damage and inhibiting NMDA receptor-mediated currents, suggesting neuroprotective potential through glutamate receptor modulation [73].

Galanthus nivalis

The main chemical component of *Galanthus nivalis*, or snowdrop, is galanthamine, an isoquinoline alkaloid. The long-acting and selective acetylcholinesterase (AChE) inhibitor galanthamine has been approved as a potentially effective treatment for AD. By allosterically modifying nicotinic acetylcholine receptors, it enhances cholinergic nicotinic neurotransmission and adds to the efficacy of treating AD [74, 75]. The medicinal potential of the Amaryllidaceae family's genus Galanthus is attributed to a variety of phytoconstituents present in its species. Two important alkaloids from this genus are used medicinally: galanthine and galantamine. Galanthamine is categorized as a reversible, competitive, and selective inhibitor of AChE. Furthermore, it improves cholinergic function and memory *via* allosterically modulating nicotinic receptors (NR) on neurons. Galantamine is used in different countries for treating AD. The NR stimulation provided by galantamine plays a crucial role in its therapeutic effects, particularly in enhancing cholinergic functions and memory. This overview describes the botanical aspects of *G. nivalis*, its distribution, phytochemistry, and detailed studies of its main constituent, galantamine, in the treatment of neurodegenerative diseases [76, 77]. *Galanthus nivalis* extract demonstrated neuroactivity by reducing anxiety-like behavior in Syrian hamsters, increasing free-roaming, object interaction, and social engagement, indicating potential benefits for neurological and psychiatric conditions [78].

Neuroprotection from Marine Sources

Since marine algae have a wide variety of structurally varied bioactive chemicals, they are known to have several positive health impacts. Studies conducted *in vivo* and *in vitro* have confirmed the anti-inflammatory and antioxidant properties of marine algae. Ecklonia cava, sometimes referred to as "sea trumpet," has been shown to have strong anti-inflammatory qualities. Pro-inflammatory mediators like prostaglandin-E2 (PGE2), nitric oxide (NO), and cytokines like interleukin-6 (IL-6), interleukin-1β (IL-1β), and tumor necrosis factor-α (TNF-α) can all be suppressed by this seaweed [86, 87].

Another sea algae called Neorhodomela articulate suppresses the production of NO and the expression of inducible nitric oxide synthase (iNOS) in BV2 cells induced by interferon-gamma (IFN-γ). Neorhodomela ocelot contains several bromophenols that have the potential to act as anti-neuroinflammatory drugs. *Laminaria japonica* yields the phytochemical fucoidan, which has a molecular weight of 7000 Daltons and is made up of 29% sulfate and 48% total sugar. Fucoidan has strong inhibitory effects on BV2 cells' generation of NO when lipopolysaccharide (LPS) is added, and it also shields rat cholinergic neurons from Aβ1-42-induced death by preventing caspase-9 and caspase-3 activation [88].

Ulva conglobata has demonstrated neuroprotective effects by inhibiting the expression of pro-inflammatory enzymes such as cyclooxygenase-2 (COX-2), which lower the synthesis of PGE2 and NO, respectively. Certain phytochemicals derived from marine algae, such as decal and phlorofucofluoroeckol, have been shown to have both inhibitory and memory-enhancing effects on acetylcholine esterase (AChE). Additionally, Ecklonia stolonifera, Hypnea valentiae, and *Ulva reticulata* showed notable impacts when different marine algae were screened for AChE inhibitory activity [89]. Due mainly to its free-radical scavenging action, bryothamnion triquetrum has shown the potential to shield GT1-7 cells from death brought on by chemical hypoxia/aglycemia. These results highlight the potential of marine algae as a source of anti-inflammatory and neuroprotective compounds.

Dietary Sources Containing Neuroprotective Activity

A diverse diet that includes a range of foods and food components is often the cause of chronic diseases such as coronary heart disease, diabetes, hypertension, and some types of cancer. However, a variety of food sources include neuroprotective properties that may lessen the likelihood of neurodegenerative illnesses. Melatonin, serotonin, and essential amino acids in the diet are essential for the health benefits of phytofoods [90]. Phytochemicals in fruits and vegetables may offer protection against disorders of the neurological system, according to epidemiological research. Fruits, vegetables, nuts, seeds, flowers, roots, bark, and other naturally occurring antioxidants include oligomeric proanthocyanidins, which have important pharmacological and therapeutic properties against oxidative stress and free radicals. As an example, proanthocyanidin extract from grape seeds offers more defense against lipid peroxidation and free radicals than vitamins C, E, and β-carotene [91]. It has been demonstrated that spirulina, spinach, and blueberries can reduce general locomotor deficits and stroke-induced cerebral infarction. A higher intake of fruits and vegetables is linked to a lower risk of Alzheimer's.

In models of Parkinson's and Alzheimer's disease, blueberries in particular have been shown to preserve dopaminergic neurons and enhance learning and memory. Research has demonstrated that pomegranate juice can enhance behavioral deficiencies, whereas apple juice can prevent cognitive deficits. Proanthocyanidins, condensed tannins, epicatechin, epicatechin gallate, catechin, and other neuroprotective substances are examples of flavonoids, which are often found in drinks including wine, tea, chocolate, and fruit juices [92].

Organosulfur compounds such as allicin and allium, which are abundant in garlic and onions, are recognized for their ability to scavenge free radicals and have neuroprotective qualities. Nuts can help stop nerve cell degeneration since they are high in omega-3 fatty acids. Walnuts are very helpful since they include monounsaturated fatty acids including oleic acid, linoleic acid, and α-linolenic acid. Resveratrol, a substance having neuroprotective qualities, is found in peanuts. Anthocyanins such as pelargonidin, cyanidin, and malvidin are abundant in berries, which include blueberries, blackberries, cherries, and strawberries. According to reports, blueberries use the ERK pathway to increase memory and reverse age-related losses in motor function by inhibiting nitric oxide, TNF-α, and IL-1β in activated microglia cells. High-fiber flours and whole grains are proven to postpone the onset of dementia [93]. Rich in tocopherols and tocotrienols, rice bran oil is a potent antioxidant that promotes neuroprotection. Furthermore to being rich in vitamins and minerals, bananas are a great source of dopamine, a neurotransmitter. Rich in vitamins B, E, and folate, green leafy vegetables help protect the brain by lowering homocysteine, an amino acid linked to the death of brain cells. Flavonones including naringenin, hesperetin, and taxifolin found in citrus fruits and tomatoes, as well as flavonoids like luteolin and apigenin found in artichokes, chives, celery, and parsley, stimulate ERK1/2 signaling in cortical neurons [94, 95].

Turmeric (*Curcuma longa*) is the source of curcumin, a powerful neuroprotective substance. It restores chronic stress-induced impairment of hippocampal neurogenesis, protects neurons against ischemic cell death and behavioral impairments in animal models, and upregulates the production of Brain-Derived Neurotrophic Factor (BDNF). In rat cortical cells, curcumin also offers protection against glutamate excitotoxicity [96].

Red grapes contain resveratrol, which has strong neuroprotective effects. It shields neurons from oxidative stress-induced mortality and ischemia injury. In models of Parkinson's disease, it also protects dopaminergic neurons from oxidative and metabolic damage. In models of Alzheimer's disease, resveratrol also helps to remove amyloid β-peptide from the body. Broccoli and Brussels sprouts contain sulforaphane, an isothiocyanate that lowers edema and brain damage while shielding neurons from mitochondrial toxins and oxidative stress [97]. Allium and allicin, two organosulfur compounds found in garlic and onions, are recognized for their neuroprotective and free radical scavenging properties. They also activate pathways linked to mitochondrial uncoupling proteins and neuroprotection. Neuroprotection is provided by caffeine and other methylxanthines, which are present in coffee, tea, and chocolate [98]. These compounds, along with other dietary phytochemicals, demonstrate the significant potential of a varied diet in supporting brain health and mitigating neurodegenerative diseases.

Recent Findings Related to Phytoconstituents for Neurodegenerative Disorder

- A study reported that the Glycation contributes to neurodegenerative disorders like Alzheimer's disease (AD) by enhancing β-amyloid (Aβ) aggregation and toxicity. Polyphenol-rich berries, including blackberry, black raspberry, blueberry, cranberry, red raspberry, and strawberry, were evaluated for their neuroprotective properties. Berry crude extracts (CE) were fractionated into anthocyanins-enriched (ACE) and anthocyanins-free (ACF) extracts. ACEs demonstrated superior free radical scavenging, reactive carbonyl species (*e.g.*, methylglyoxal) trapping, and anti-glycation activities compared to ACFs. They also inhibited thermal- and methylglyoxal-induced Aβ fibrillation and reduced oxidative stress, nitric oxide production, and cytotoxicity in BV-2 microglia. These findings highlight the potential of berry ACEs in mitigating glycation-associated damage and Aβ toxicity [99].

- Sesame oil (SO), derived from Sesamum indicum, has traditional medicinal use and notable anti-inflammatory and antioxidant properties. This study explored its neuroprotective effects against Alzheimer's disease (AD)-like symptoms induced by aluminum chloride (AlCl3) in rats. Over six weeks, rats treated with AlCl3 (100 mg/kg) exhibited cognitive impairments, oxidative stress, and neuroinflammation. Behavioral assessments, such as the Morris water maze, revealed significant improvements in learning and memory in SO-treated groups. SO reduced elevated acetylcholinesterase (AChE) and amyloid beta (Aβ) levels while mitigating AlCl3-induced histopathological brain changes. SO also decreased inflammatory markers, including tumor necrosis factor-alpha (TNF-α) and interleukin-1 beta (IL-1β), and restored oxidative stress balance. Additionally, SO inhibited p38 mitogen-activated protein kinase (p38MAPK) activation, increased brain-derived neurotrophic factor (BDNF), and modulated peroxisome proliferator-activated receptor gamma (PPAR-γ) and nuclear factor kappa B (NF-κB) expression. These findings suggest that SO's neuroprotective effects result from its modulation of oxidative stress, neuroinflammation, and cognitive functions, potentially involving NF-κB/p38MAPK/BDNF/PPAR-γ signalling pathways [100].

Natural herbs, in whole form or as extracts, show promise as neuroprotective agents for managing neurodegenerative disorders (NDs). Key phytoconstituents such as resveratrol, turmeric, and apigenin have demonstrated neuroprotective properties, including antioxidative, anti-inflammatory, and anti-amyloidogenic effects. However, their clinical efficacy is hindered by poor bioavailability, limited physicochemical stability, solubility issues, and challenges in crossing the Blood-Brain Barrier (BBB). Future research should prioritize rigorous, large-scale clinical trials and standardized methodologies to optimize safety profiles, therapeutic efficacy, and dosage guidelines for phytoconstituents in ND treatment.

CONCLUSION

Neurodegenerative disorders are characterized by the gradual death of brain cells, driven by complex pathophysiological mechanisms such as oxidative stress, inflammation, excitotoxicity, and imbalances in neurotransmitters. The interplay of Reactive Nitrogen Species (RNS), Reactive Oxygen Species (ROS), and neuroinflammatory processes underscores the intricate nature of these conditions and highlights the need for multifaceted therapeutic approaches. Medicinal plants such as *Echinacea purpurea, Matricaria chamomilla, St. John's Wort, Boswellia serrata, Lepidium meyenii, Acorus calamus, Vitis vinifera, Celastrus paniculatus, Hyoscyamus niger,* and *Galanthus nivalis* offer promising neuroprotective potential due to their anti-inflammatory, antioxidant, neuroprotective, and cognitive-enhancing properties. These natural compounds target multiple pathways, presenting a holistic strategy for improving cognitive function and mitigating neurodegeneration. Furthermore, incorporating marine phytochemicals and bioactives into therapeutic regimens may complement these effects. To maximize the impact of these findings, future research should focus on elucidating the precise molecular mechanisms of action, standardizing extraction methods, optimizing dosages, and conducting rigorous preclinical and clinical trials. Establishing efficacy and safety profiles will provide critical insights for integrating these agents into healthcare practices, ultimately benefiting patients and healthcare providers alike. This multi-pronged approach will ensure the translation of natural neuroprotective agents from research to real-world application.

REFERENCES

[1]   Verkhratsky, A.; Ho, M.S.; Zorec, R.; Parpura, V. *The concept of neuroglia. Adv. Exp. Med. Biol*; NIH Public Access, **2019**, 1175, pp. 1-13.
        [http://dx.doi.org/10.1007/978-981-13-9913-8_1]

[2]   Damisah, E.C.; Hill, R.A.; Rai, A.; Chen, F.; Rothlin, C.V.; Ghosh, S.; Grutzendler, J. Astrocytes and microglia play orchestrated roles and respect phagocytic territories during neuronal corpse removal *in vivo. Sci. Adv.,* **2020**, *6*(26), eaba3239.
        [http://dx.doi.org/10.1126/sciadv.aba3239] [PMID: 32637606]

[3]   Verkhratsky, A.; Parpura, V. Neurological and psychiatric disorders as a neuroglial failure. *Period. Biol.,* **2014**, *116*(2), 115-124.
        [PMID: 25544781]

[4]   Kumar, G.P.; Khanum, F. Neuroprotective potential of phytochemicals. *Pharmacogn. Rev.,* **2012**, *6*(12), 81-90.
        [http://dx.doi.org/10.4103/0973-7847.99898] [PMID: 23055633]

[5]   Dugger, B.N.; Dickson, D.W. Pathology of neurodegenerative diseases. *Cold Spring Harb. Perspect. Biol.,* **2017**, *9*(7), a028035.

[http://dx.doi.org/10.1101/cshperspect.a028035] [PMID: 28062563]

[6]   Kantawala, B.; Ramadan, N.; Hassan, Y.; Fawaz, V.; Mugisha, N.; Nazir, A.; Wojtara, M.; Uwishema, O. Physical activity intervention for the prevention of neurological diseases. *Health Sci. Rep.,* **2023**, *6*(8), e1524.
[http://dx.doi.org/10.1002/hsr2.1524] [PMID: 37614284]

[7]   Karimi, A.; Majlesi, M.; Rafieian-Kopaei, M. Herbal versus synthetic drugs; beliefs and facts. *J. Nephropharmacol.,* **2015**, *4*(1), 27-30.
[PMID: 28197471]

[8]   Atanasov, A.G.; Waltenberger, B.; Pferschy-Wenzig, E.M.; Linder, T.; Wawrosch, C.; Uhrin, P.; Temml, V.; Wang, L.; Schwaiger, S.; Heiss, E.H.; Rollinger, J.M.; Schuster, D.; Breuss, J.M.; Bochkov, V.; Mihovilovic, M.D.; Kopp, B.; Bauer, R.; Dirsch, V.M.; Stuppner, H. Discovery and resupply of pharmacologically active plant-derived natural products: A review. *Biotechnol. Adv.,* **2015**, *33*(8), 1582-1614.
[http://dx.doi.org/10.1016/j.biotechadv.2015.08.001] [PMID: 26281720]

[9]   Katiyar, C.; Kanjilal, S.; Gupta, A.; Katiyar, S. Drug discovery from plant sources: An integrated approach. *Ayu,* **2012**, *33*(1), 10-19.
[http://dx.doi.org/10.4103/0974-8520.100295] [PMID: 23049178]

[10]   Moharana, A.; Choudhury, P.; Behera, S.R.; Vishwakarma, P.K.; Tripathi, S.M.; Srivastava, S.P. Exploring the Pharmacological Potential of Lobelia trigona and its Bioactive Compounds. *Current Nutraceuticals,* **2024**, *5*, e040324227605.
[http://dx.doi.org/10.2174/0126659786278190240214062949]

[11]   Lima, A.A.; Mridha, M.F.; Das, S.C.; Kabir, M.M.; Islam, M.R.; Watanobe, Y. A Comprehensive Survey on the Detection, Classification, and Challenges of Neurological Disorders. *Biology (Basel),* **2022**, *11*(3), 469.
[http://dx.doi.org/10.3390/biology11030469] [PMID: 35336842]

[12]   Penninx, B.W.J.H.; Pine, D.S.; Holmes, E.A.; Reif, A. Anxiety disorders. *Lancet,* **2021**, *397*(10277), 914-927.
[http://dx.doi.org/10.1016/S0140-6736(21)00359-7]

[13]   Deb, S.S.; Perera, B.; Bertelli, M.O. *Attention Deficit Hyperactivity Disorder*; Textb. Psychiatry Intellect. Disabil. Autism Spectr. Disord, **2022**, pp. 457-482.
[http://dx.doi.org/10.1007/978-3-319-95720-3_17]

[14]   Al-abri, K.; Edge, D.; Armitage, C.J. Prevalence and correlates of perinatal depression. *Soc. Psychiatry Psychiatr. Epidemiol.,* **2023**, *58*(11), 1581-1590.
[http://dx.doi.org/10.1007/s00127-022-02386-9] [PMID: 36646936]

[15]   Chowdhury, M.W.A.; Ahmed, H.U.; Hossain, M.D.; Aftab, A.; Soron, T.R.; Alam, M.T.; Uddin, A. Suicide and depression in the World Health Organization South-East Asia Region: A systematic review. *WHO South-East Asia J. Public Health,* **2017**, *6*(1), 60-66.
[http://dx.doi.org/10.4103/2224-3151.206167] [PMID: 28597861]

[16]   Mikulska, J.; Juszczyk, G.; Gawrońska-Grzywacz, M.; Herbet, M. Hpa axis in the pathomechanism of depression and schizophrenia: New therapeutic strategies based on its participation. *Brain Sci.,* **2021**, *11*(10), 1298.
[http://dx.doi.org/10.3390/brainsci11101298] [PMID: 34679364]

[17]   Błaszczyk, J.W. Parkinson's disease and neurodegeneration: GABA-collapse hypothesis. *Front. Neurosci.,* **2016**, *10*, 269.
[http://dx.doi.org/10.3389/fnins.2016.00269] [PMID: 27375426]

[18]   Hardy, J.A.; Higgins, G.A. Alzheimer's disease: The amyloid cascade hypothesis. *Science,* **1992**, *256*(5054), 184–185.
[http://dx.doi.org/10.1126/science.1566067]

[19]   Stafstrom, C.E.; Carmant, L. Seizures and epilepsy: an overview for neuroscientists. *Cold Spring Harb. Perspect. Med.,* **2015**, *5*(6), a022426.
[http://dx.doi.org/10.1101/cshperspect.a022426] [PMID: 26033084]

[20]   Neves, D.; Salazar, I.L.; Almeida, R.D.; Silva, R.M. Molecular mechanisms of ischemia and glutamate excitotoxicity. *Life Sci.,* **2023**, *328*, 121814.
[http://dx.doi.org/10.1016/j.lfs.2023.121814] [PMID: 37236602]

[21]   Stępnicki, P.; Kondej, M.; Kaczor, A.A. Current concepts and treatments of schizophrenia. *Molecules,* **2018**, *23*(8), 2087.
[http://dx.doi.org/10.3390/molecules23082087] [PMID: 30127324]

[22]   Bathina, S.; Das, U.N. Brain-derived neurotrophic factor and its clinical implications. *Arch. Med. Sci.,* **2015**, *6*(6), 1164-1178.
[http://dx.doi.org/10.5114/aoms.2015.56342] [PMID: 26788077]

[23]   Libro, R.; Giacoppo, S.; Soundara Rajan, T.; Bramanti, P.; Mazzon, E. Natural phytochemicals in the treatment and prevention of dementia: An overview. *Molecules,* **2016**, *21*(4), 518.
[http://dx.doi.org/10.3390/molecules21040518] [PMID: 27110749]

[24]   Williams, M.; Porsolt, R.D. *CNS safety pharmacology. xPharm Compr. Pharmacol. Ref;* Elsevier Inc., **2007**, pp. 1-13.
[http://dx.doi.org/10.1016/B978-008055232-3.63682-7]

[25]   Lee, K.H.; Cha, M.; Lee, B.H. Neuroprotective effect of antioxidants in the brain. *Int. J. Mol. Sci.,* **2020**, *21*(19), 7152.
[http://dx.doi.org/10.3390/ijms21197152] [PMID: 32998277]

[26]   Jiang, M.; Jang, S.E.; Zeng, L. The Effects of Extrinsic and Intrinsic Factors on Neurogenesis. *Cells,* **2023**, *12*(9), 1285.
[http://dx.doi.org/10.3390/cells12091285] [PMID: 37174685]

[27]   Md, S.; Alhakamy, N.A.; Aldawsari, H.M.; Asfour, H.Z. Neuroprotective and antioxidant effect of naringenin-loaded nanoparticles for nose-to-brain delivery. *Brain Sci.,* **2019**, *9*(10), 275.
[http://dx.doi.org/10.3390/brainsci9100275] [PMID: 31618942]

[28]   Stephenson, J.; Nutma, E.; van der Valk, P.; Amor, S. Inflammation in CNS neurodegenerative diseases. *Immunology,* **2018**, *154*(2), 204-219.
[http://dx.doi.org/10.1111/imm.12922] [PMID: 29513402]

[29]   Zhao, X.Y.; Lu, M.H.; Yuan, D.J.; Xu, D.E.; Yao, P.P.; Ji, W.L.; Chen, H.; Liu, W.L.; Yan, C.X.; Xia, Y.Y.; Li, S.; Tao, J.; Ma, Q.H. Mitochondrial dysfunction in neural injury. *Front. Neurosci.,* **2019**, *13*, 30.
[http://dx.doi.org/10.3389/fnins.2019.00030] [PMID: 30778282]

[30]   Arundine, M.; Tymianski, M. Molecular mechanisms of glutamate-dependent neurodegeneration in ischemia and traumatic brain injury. *Cell. Mol. Life Sci.,* **2004**, *61*(6), 657-668.
[http://dx.doi.org/10.1007/s00018-003-3319-x] [PMID: 15052409]

[31]   Burlou-Nagy, C.; Bănică, F.; Jurca, T.; Vicaş, L.G.; Marian, E.; Muresan, M.E.; Bácskay, I.; Kiss, R.; Fehér, P.; Pallag, A. *Echinacea purpurea* (L.) Moench: Biological and Pharmacological Properties. A Review. *Plants,* **2022**, *11*(9), 1244.
[http://dx.doi.org/10.3390/plants11091244] [PMID: 35567246]

[32]   Zdor, V.N.; Pospelov, S.V. BIOLOGICAL ACTIVITY OF ECHINACEA PURPURE EXTRACTS (ECHINACEA PURPUREA (L.) MOENCH.). *National Association of Scientists,* **2020**, *1*(29(56)), 31-35.
[http://dx.doi.org/10.31618/nas.2413-5291.2020.1.56.235]

[33]   Abdel-Wahhab, K.G.; Sayed, R.S.; EL-Sahra, D.G.; Hassan, L.K.; Elqattan, G.M.; Mannaa, F.A. *Echinacea purpurea* extract intervention for counteracting neurochemical and behavioral changes induced by bifenthrin. *Metab. Brain Dis.,* **2023**, *39*(1), 101-113.

[http://dx.doi.org/10.1007/s11011-023-01303-6] [PMID: 38150137]

[34] Oniszczuk, T.; Oniszczuk, A.; Gondek, E.; Guz, L.; Puk, K.; Kocira, A.; Kusz, A.; Kasprzak, K.; Wójtowicz, A. Active polyphenolic compounds, nutrient contents and antioxidant capacity of extruded fish feed containing purple coneflower (*Echinacea purpurea* (L.) Moench.). *Saudi J. Biol. Sci.,* **2019**, *26*(1), 24-30.
[http://dx.doi.org/10.1016/j.sjbs.2016.11.013] [PMID: 30622403]

[35] Alshehri, A.A.; Malik, M.A. Phytomediated photo-induced green synthesis of silver nanoparticles using *Matricaria chamomilla* L. and its catalytic activity against rhodamine B. *Biomolecules,* **2020**, *10*(12), 1604.
[http://dx.doi.org/10.3390/biom10121604] [PMID: 33256218]

[36] Akram, W.; Ahmed, S.; Rihan, M.; Arora, S.; Khalid, M.; Ahmad, S.; Ahmad, F.; Haque, S.; Vashishth, R. An updated comprehensive review of the therapeutic properties of Chamomile (*Matricaria chamomilla* L.). *Int. J. Food Prop.,* **2024**, *27*(1), 133-164.
[http://dx.doi.org/10.1080/10942912.2023.2293661]

[37] El Mihyaoui, A.; Esteves da Silva, J.C.G.; Charfi, S.; Candela Castillo, M.E.; Lamarti, A.; Arnao, M.B. Chamomile (*Matricaria chamomilla* L.): A Review of Ethnomedicinal Use, Phytochemistry and Pharmacological Uses. *Life (Basel),* **2022**, *12*(4), 479.
[http://dx.doi.org/10.3390/life12040479] [PMID: 35454969]

[38] Sah, A.; Naseef, P.P.; Kuruniyan, M.S.; Jain, G.K.; Zakir, F.; Aggarwal, G. A Comprehensive Study of Therapeutic Applications of Chamomile. *Pharmaceuticals (Basel),* **2022**, *15*(10), 1284.
[http://dx.doi.org/10.3390/ph15101284] [PMID: 36297396]

[39] Moshfegh, A.; Setorki, M. Neuroprotective Effect of *Matricaria chamomilla* Extract on Motor Dysfunction Induced by Transient Global Cerebral Ischemia and Reperfusion in Rat. *Zahedan J. Res. Med. Sci.,* **2017**, *19*(9)
[http://dx.doi.org/10.5812/zjrms.10927]

[40] Yilmazoğlu, E.; Hasdemïr, M.; Hasdemïr, B. Recent Studies on Antioxidant, Antimicrobial, and Ethnobotanical Uses of Hypericum perforatum L. (Hypericaceae). *Journal of the Turkish Chemical Society Section A: Chemistry,* **2022**, *9*(2), 373-394.
[http://dx.doi.org/10.18596/jotcsa.1024791]

[41] Mohamed, F.F.; Anhlan, D.; Schöfbänker, M.; Schreiber, A.; Classen, N.; Hensel, A.; Hempel, G.; Scholz, W.; Kühn, J.; Hrincius, E.R.; Ludwig, S. *Hypericum perforatum* and Its Ingredients Hypericin and Pseudohypericin Demonstrate an Antiviral Activity against SARS-CoV-2. *Pharmaceuticals (Basel),* **2022**, *15*(5), 530.
[http://dx.doi.org/10.3390/ph15050530] [PMID: 35631357]

[42] Shakya, P.; Marslin, G.; Siram, K.; Beerhues, L.; Franklin, G. Elicitation as a tool to improve the profiles of high-value secondary metabolites and pharmacological properties of *Hypericum perforatum*. *J. Pharm. Pharmacol.,* **2019**, *71*(1), 70-82.
[http://dx.doi.org/10.1111/jphp.12743] [PMID: 28523644]

[43] Oliveira, A.I.; Pinho, C.; Sarmento, B.; Dias, A.C.P. Neuroprotective activity of hypericum perforatum and its major components. *Front. Plant Sci.,* **2016**, *7*, 1004.
[http://dx.doi.org/10.3389/fpls.2016.01004] [PMID: 27462333]

[44] Behnke, K.; Jensen, G.S.; Graubaum, H.J.; Gruenwald, J. Hypericum perforatum versus fluoxetine in the treatment of mild to moderate depression. *Adv. Ther.,* **2002**, *19*(1), 43-52.
[http://dx.doi.org/10.1007/BF02850017] [PMID: 12008860]

[45] Siddiqui, M.Z. Boswellia serrata, a potential antiinflammatory agent: an overview. *Indian J. Pharm. Sci.,* **2011**, *73*(3), 255-261.
[http://dx.doi.org/10.4103/0250-474X.93507] [PMID: 22457547]

[46] Zapata, A.; Fernández-Parra, R. Management of Osteoarthritis and Joint Support Using Feed Supplements: A Scoping Review of Undenatured Type II Collagen and *Boswellia serrata*. *Animals*

(Basel), **2023**, *13*(5), 870.
[http://dx.doi.org/10.3390/ani13050870] [PMID: 36899726]

[47]   Marefati, N.; Beheshti, F.; Vafaee, F.; Barabadi, M.; Hosseini, M. The Effects of Incensole Acetate on Neuro-inflammation, Brain-Derived Neurotrophic Factor and Memory Impairment Induced by Lipopolysaccharide in Rats. *Neurochem. Res.,* **2021**, *46*(9), 2473-2484.
[http://dx.doi.org/10.1007/s11064-021-03381-3] [PMID: 34173963]

[48]   Mahboubi, M.; Kashani, L.M.T. Boswellia serrata Oleo-Gum-Resin and its Effect on Memory Functions: A Review. *Nat. Prod. J.,* **2020**, *10*(4), 355-363.
[http://dx.doi.org/10.2174/2210315509666190311153819]

[49]   Kumar, V.; Sharma, C.; Taleuzzaman, M.; Nagarajan, K.; Haque, A.; Bhatia, M.; Khan, S.; Salkini, M.A.; Bhatt, P. Neuroprotective Effect of *Boswellia serrata* against 3-NP Induced Experimental Huntington's Disease. *Curr. Bioact. Compd.,* **2024**, *20*(6), e180124225809.
[http://dx.doi.org/10.2174/0115734072272233231119161319]

[50]   Yan, S.; Wei, J.; Chen, R. Evaluation of the Biological Activity of Glucosinolates and Their Enzymolysis Products Obtained from *Lepidium meyenii Walp* (Maca). *Int. J. Mol. Sci.,* **2022**, *23*(23), 14756.
[http://dx.doi.org/10.3390/ijms232314756] [PMID: 36499083]

[51]   Ulloa del Carpio, N.; Alvarado-Corella, D.; Quiñones-Laveriano, D.M.; Araya-Sibaja, A.; Vega-Baudrit, J.; Monagas-Juan, M.; Navarro-Hoyos, M.; Villar-López, M. Exploring the chemical and pharmacological variability of *Lepidium meyenii*: a comprehensive review of the effects of maca. *Front. Pharmacol.,* **2024**, *15*, 1360422.
[http://dx.doi.org/10.3389/fphar.2024.1360422] [PMID: 38440178]

[52]   Guo, S.S.; Gao, X.F.; Gu, Y.R.; Wan, Z.X.; Lu, A.M.; Qin, Z.H.; Luo, L. Preservation of Cognitive Function by *Lepidium meyenii* (Maca) Is Associated with Improvement of Mitochondrial Activity and Upregulation of Autophagy-Related Proteins in Middle-Aged Mouse Cortex. *Evid. Based Complement. Alternat. Med.,* **2016**, *2016*(1), 4394261.
[http://dx.doi.org/10.1155/2016/4394261] [PMID: 27648102]

[53]   Paithankar, V.V.; Belsare, S.L.; Charde, R.M.; Vyas, J.V. Acorus calamus: An overview. *Int. J. Biomed. Res.,* **2011**, *2*(10), 2740-2745.
[http://dx.doi.org/10.7439/ijbr.v2i10.174]

[54]   Puthur, S.; Raj, K.K.; Anoopkumar, A.N.; Rebello, S.; Aneesh, E.M. *Acorus calamus* mediated green synthesis of ZnONPs: A novel nano antioxidant to future perspective. *Adv. Powder Technol.,* **2020**, *31*(12), 4679-4682.
[http://dx.doi.org/10.1016/j.apt.2020.10.016]

[55]   Rajput, S.B.; Tonge, M.B.; Karuppayil, S.M. An overview on traditional uses and pharmacological profile of *Acorus calamus* Linn. (Sweet flag) and other Acorus species. *Phytomedicine,* **2014**, *21*(3), 268-276.
[http://dx.doi.org/10.1016/j.phymed.2013.09.020] [PMID: 24200497]

[56]   Sharma, V.; Sharma, R.; Gautam, D.; Kuca, K.; Nepovimova, E.; Martins, N. Role of vacha (*Acorus calamus* Linn.) in neurological and metabolic disorders: Evidence from ethnopharmacology, phytochemistry, pharmacology and clinical study. *J. Clin. Med.,* **2020**, *9*(4), 1176.
[http://dx.doi.org/10.3390/jcm9041176] [PMID: 32325895]

[57]   Mamta, S.; Jyoti, S. Phytochemical Screening of Acorus Calamus and Lantana Camara. *Int Res J Pharm,* **2012**, *3*, 324-326.

[58]   Malik, R.; Kalra, S. Neurobiology of depression: Insights and therapeutic implications. *Brain Research,* **2024**, *1822*, 148616.
[http://dx.doi.org/10.1016/j.brainres.2023.148616]

[59]   Ingle, S.T.; Srivastava, J.N.; Shete, R.S. *Diseases of Grapevine (Vitis Vinifera L.) and Their Management*; Dis. Hortic. Crop, **2022**, pp. 201-216.

[http://dx.doi.org/10.1201/9781003160397-11]

[60] Ammar A, J. Kushaiba R, A. Harakat S, A. Alennabi K. Antioxidant and Antibacterial Effect of *Vitis labrusca, Vitis vinifera* and *Vitis vinifera* Seeds Extract. *South Asian Res J Pharm Sci,* **2021**, *3*, 34-39. [http://dx.doi.org/10.36346/sarjps.2021.v03i02.003]

[61] Jadidian, F.; Amirhosseini, M.; Abbasi, M.; Hamedanchi, N.F.; Zerangian, N.; Erabi, G. Pharmacotherapeutic potential of *Vitis vinifera* (grape) in age-related neurological diseases. *Bol. Latinoam. Caribe Plantas Med. Aromat.,* **2024**, *23*(3), 349-370. [http://dx.doi.org/10.37360/blacpma.24.23.3.24]

[62] Pazos-Tomas, C.C.; Cruz-Venegas, A.; Pérez-Santiago, A.D.; Sánchez-Medina, M.A.; Matías-Pérez, D.; García-Montalvo, I.A. *Vitis vinifera*: An alternative for the prevention of neurodegenerative diseases. *J. Oleo Sci.,* **2020**, *69*(10), 1147-1161. [http://dx.doi.org/10.5650/jos.ess20109] [PMID: 32908097]

[63] Rapaka, D.; Bitra, V.R.; Vishala, T.C.; Akula, A. *Vitis vinifera* acts as anti-Alzheimer's agent by modulating biochemical parameters implicated in cognition and memory. *J. Ayurveda Integr. Med.,* **2019**, *10*(4), 241-247. [http://dx.doi.org/10.1016/j.jaim.2017.06.013] [PMID: 30337026]

[64] Choi, J.; Choi, S.Y.; Hong, Y.; Han, Y.E.; Oh, S.J.; Lee, B.; Choi, C.W.; Kim, M.S. The central administration of vitisin a, extracted from Vitis vinifera, improves cognitive function and related signaling pathways in a scopolamine-induced dementia model. *Biomed. Pharmacother.,* **2023**, *163*, 114812. [http://dx.doi.org/10.1016/j.biopha.2023.114812] [PMID: 37148861]

[65] Inglot, J.; Aebisher, D.; Bartusik-Aebisher, D. *Hyoscyamus niger*; Biochem. Guid. to Toxins, **2023**, pp. 61-64. [http://dx.doi.org/10.1055/b-0036-138552]

[66] Teut, M. Homeopathic treatment of patients with dementia. *Am J Homeopath Med,* **2010**, *103*, 120-124.

[67] Shim, K.H.; Kang, M.J.; Sharma, N.; An, S.S.A. Beauty of the beast: anticholinergic tropane alkaloids in therapeutics. *Nat. Prod. Bioprospect.,* **2022**, *12*(1), 33. [http://dx.doi.org/10.1007/s13659-022-00357-w] [PMID: 36109439]

[68] Sengupta, T.; Vinayagam, J.; Nagashayana, N.; Gowda, B.; Jaisankar, P.; Mohanakumar, K.P. Antiparkinsonian effects of aqueous methanolic extract of Hyoscyamus niger seeds result from its monoamine oxidase inhibitory and hydroxyl radical scavenging potency. *Neurochem. Res.,* **2011**, *36*(1), 177-186. [http://dx.doi.org/10.1007/s11064-010-0289-x] [PMID: 20972705]

[69] Nagpal, K.; Garg, M.; Arora, D.; Dubey, A.; Grewal, A.S. An extensive review on phytochemistry and pharmacological activities of Indian medicinal plant *Celastrus paniculatus* Willd. *Phytother. Res.,* **2022**, *36*(5), 1930-1951. [http://dx.doi.org/10.1002/ptr.7424] [PMID: 35199395]

[70] Parimala, S.; Shashidhar, G.; Sridevi, C.; Jyothi, V.; Suthakaran, R. Anti- inflammatory activity of Celastrus paniculatus seeds. *Int. J. Pharm. Tech. Res.,* **2009**, *1*, 1326-1329.

[71] Faldu, K.G.; Patel, S.S.; Shah, J.S. Celastrus paniculatus oil ameliorates synaptic plasticity in a rat model of attention deficit hyperactivity disorder. *Asian Pac. J. Trop. Biomed.,* **2021**, *11*(3), 105-114. [http://dx.doi.org/10.4103/2221-1691.306690]

[72] Sankaramourthy, D.; Sankaranarayanan, L.; Subramanian, K.; Sadras, S.R. Neuroprotective potential of *Celastrus paniculatus* seeds against common neurological ailments: a narrative review. *J. Complement. Integr. Med.,* **2023**, *20*(3), 530-536. [http://dx.doi.org/10.1515/jcim-2021-0448] [PMID: 35005853]

[73] Godkar, P.B.; Gordon, R.K.; Ravindran, A.; Doctor, B.P. Celastrus paniculatus seed water soluble

extracts protect against glutamate toxicity in neuronal cultures from rat forebrain. *J. Ethnopharmacol.,* **2004**, *93*(2-3), 213-219.
[http://dx.doi.org/10.1016/j.jep.2004.03.051] [PMID: 15234755]

[74]   Mantle, D.; Pickering, A.T.; Perry, E.K. Medicinal plant extracts for the treatment of dementia: A review of their pharmacology, efficacy and tolerability. *CNS Drugs,* **2000**, *13*(3), 201-213.
[http://dx.doi.org/10.2165/00023210-200013030-00006]

[75]   Berkov, S.; Codina, C.; Viladomat, F.; Bastida, J. Alkaloids from Galanthus nivalis. *Phytochemistry,* **2007**, *68*(13), 1791-1798.
[http://dx.doi.org/10.1016/j.phytochem.2007.03.025] [PMID: 17475295]

[76]   Howes, M.J.R.; Perry, N.S.L.; Houghton, P.J. Plants with traditional uses and activities, relevant to the management of Alzheimer's disease and other cognitive disorders. *Phytother. Res.,* **2003**, *17*(1), 1-18.
[http://dx.doi.org/10.1002/ptr.1280] [PMID: 12557240]

[77]   Sagrario, M.A.; Mp, G.; J, B.; P, B-B. Plants with evidence-based therapeutic effects against neurodegenerative diseases. *Pharm. Pharmacol. Int. J.,* **2019**, *7*(5), 221-227.
[http://dx.doi.org/10.15406/ppij.2019.07.00255]

[78]   Kip, H.A.; Kuro, M.; Hami, S.; Nugget, T.; Montre, E.R. Galanthum nivalis Extract is Neurologically Active and Improves Anxiety and Social Interactions in *Mesocricetus auratus. Nat. Prod. J.,* **2019**, *9*(2), 133-137.
[http://dx.doi.org/10.2174/2210315508666180515100620]

[79]   Moretti, M.; Rodrigues, A.L.S. Functional role of ascorbic acid in the central nervous system: a focus on neurogenic and synaptogenic processes. *Nutr. Neurosci.,* **2022**, *25*(11), 2431-2441.
[http://dx.doi.org/10.1080/1028415X.2021.1956848] [PMID: 34493165]

[80]   Kim, S.M.; Park, Y.J.; Shin, M.S.; Kim, H.R.; Kim, M.J.; Lee, S.H.; Yun, S.P.; Kwon, S.H. Acacetin inhibits neuronal cell death induced by 6-hydroxydopamine in cellular Parkinson's disease model. *Bioorg. Med. Chem. Lett.,* **2017**, *27*(23), 5207-5212.
[http://dx.doi.org/10.1016/j.bmcl.2017.10.048] [PMID: 29089232]

[81]   Dey, A.; Bhattacharya, R.; Mukherjee, A.; Pandey, D.K. Natural products against Alzheimer's disease: Pharmaco-therapeutics and biotechnological interventions. *Biotechnol. Adv.,* **2017**, *35*(2), 178-216.
[http://dx.doi.org/10.1016/j.biotechadv.2016.12.005] [PMID: 28043897]

[82]   de Andrade Teles, R.B.; Diniz, T.C.; Costa Pinto, T.C.; de Oliveira Júnior, R.G.; Gama e Silva, M.; de Lavor, É.M.; Fernandes, A.W.C.; de Oliveira, A.P.; de Almeida Ribeiro, F.P.R.; da Silva, A.A.M.; Cavalcante, T.C.F.; Quintans Júnior, L.J.; da Silva Almeida, J.R.G. Flavonoids as therapeutic agents in Alzheimer's and Parkinson's diseases: A systematic review of preclinical evidences. *Oxid. Med. Cell. Longev.,* **2018**, *2018*(1), 7043213.
[http://dx.doi.org/10.1155/2018/7043213] [PMID: 29861833]

[83]   Wightman, E.L. Potential benefits of phytochemicals against Alzheimer's disease. *Proc. Nutr. Soc.,* **2017**, *76*(2), 106-112.
[http://dx.doi.org/10.1017/S0029665116002962] [PMID: 28143625]

[84]   An, Y.W.; Jhang, K.A.; Woo, S.Y.; Kang, J.L.; Chong, Y.H. Sulforaphane exerts its anti-inflammatory effect against amyloid-β peptide *via* STAT-1 dephosphorylation and activation of Nrf2/HO-1 cascade in human THP-1 macrophages. *Neurobiol. Aging,* **2016**, *38*, 1-10.
[http://dx.doi.org/10.1016/j.neurobiolaging.2015.10.016] [PMID: 26827637]

[85]   Wang, M.; Li, Y.J.; Ding, Y.; Zhang, H.N.; Sun, T.; Zhang, K.; Yang, L.; Guo, Y.Y.; Liu, S.B.; Zhao, M.G.; Wu, Y.M. Silibinin Prevents Autophagic Cell Death upon Oxidative Stress in Cortical Neurons and Cerebral Ischemia-Reperfusion Injury. *Mol. Neurobiol.,* **2016**, *53*(2), 932-943.
[http://dx.doi.org/10.1007/s12035-014-9062-5] [PMID: 25561437]

[86]   Ghallab, D.S.; Ibrahim, R.S.; Mohyeldin, M.M.; Shawky, E. Marine algae: A treasure trove of bioactive anti-inflammatory compounds. *Mar. Pollut. Bull.,* **2024**, *199*, 116023.
[http://dx.doi.org/10.1016/j.marpolbul.2023.116023] [PMID: 38211540]

[87] Barbalace, M.C.; Malaguti, M.; Giusti, L.; Lucacchini, A.; Hrelia, S.; Angeloni, C. Anti-inflammatory activities of marine algae in neurodegenerative diseases. *Int. J. Mol. Sci.,* **2019**, *20*(12), 3061.
[http://dx.doi.org/10.3390/ijms20123061] [PMID: 31234555]

[88] Svensson, C.; Fernaeus, S.Z.; Part, K.; Reis, K.; Land, T. LPS-induced iNOS expression in Bv-2 cells is suppressed by an oxidative mechanism acting on the JNK pathway—A potential role for neuroprotection. *Brain Res.,* **2010**, *1322*, 1-7.
[http://dx.doi.org/10.1016/j.brainres.2010.01.082] [PMID: 20138851]

[89] Jin, D.Q.; Lim, C.S.; Sung, J.Y.; Choi, H.G.; Ha, I.; Han, J.S. Ulva conglobata, a marine algae, has neuroprotective and anti-inflammatory effects in murine hippocampal and microglial cells. *Neurosci. Lett.,* **2006**, *402*(1-2), 154-158.
[http://dx.doi.org/10.1016/j.neulet.2006.03.068] [PMID: 16644126]

[90] Keservani, RK; Kesharwani, RK; Emerald, M; Sharma, AK Nutraceutical fruits and foods for neurodegenerative disorders. *Academic Press,* **2023**.
[http://dx.doi.org/10.1016/C2022-0-00110-0]

[91] Kumar, A.; P, N.; Kumar, M.; Jose, A.; Tomer, V.; Oz, E.; Proestos, C.; Zeng, M.; Elobeid, T.; K, S.; Oz, F. Major Phytochemicals: Recent Advances in Health Benefits and Extraction Method. *Molecules,* **2023**, *28*(2), 887.
[http://dx.doi.org/10.3390/molecules28020887] [PMID: 36677944]

[92] Essa, M.M.; Subash, S.; Al-Adawi, S.; Memon, M.; Manivasagam, T.; Akbar, M. Neuroprotective effects of berry fruits on neurodegenerative diseases. *Neural Regen. Res.,* **2014**, *9*(16), 1557-1566.
[http://dx.doi.org/10.4103/1673-5374.139483] [PMID: 25317174]

[93] Jikah, A.N.; Edo, G.I. Mechanisms of action by sulphur compounds in Allium sativum. A review. *Pharmacol. Res. Mod. Chin. Med.,* **2023**, *9*, 100323.
[http://dx.doi.org/10.1016/j.prmcm.2023.100323]

[94] Gómez-Pinilla, F. Brain foods: the effects of nutrients on brain function. *Nat. Rev. Neurosci.,* **2008**, *9*(7), 568-578.
[http://dx.doi.org/10.1038/nrn2421] [PMID: 18568016]

[95] Lubis, A.; Wakiah, N. Nutrition and Psychiatric Disorders. *J. Ment. Health,* **2023**, *32*(2), 534-534.
[http://dx.doi.org/10.1080/09638237.2023.2182434]

[96] Xu, Y.; Ku, B.; Cui, L.; Li, X.; Barish, P.A.; Foster, T.C.; Ogle, W.O. Curcumin reverses impaired hippocampal neurogenesis and increases serotonin receptor 1A mRNA and brain-derived neurotrophic factor expression in chronically stressed rats. *Brain Res.,* **2007**, *1162*, 9-18.
[http://dx.doi.org/10.1016/j.brainres.2007.05.071] [PMID: 17617388]

[97] Rahman, M.H.; Akter, R.; Bhattacharya, T.; Abdel-Daim, M.M.; Alkahtani, S.; Arafah, M.W.; Al-Johani, N.S.; Alhoshani, N.M.; Alkeraishan, N.; Alhenaky, A.; Abd-Elkader, O.H.; El-Seedi, H.R.; Kaushik, D.; Mittal, V. Resveratrol and Neuroprotection: Impact and Its Therapeutic Potential in Alzheimer's Disease. *Front. Pharmacol.,* **2020**, *11*, 619024.
[http://dx.doi.org/10.3389/fphar.2020.619024] [PMID: 33456444]

[98] Shang, A.; Cao, S.Y.; Xu, X.Y.; Gan, R.Y.; Tang, G.Y.; Corke, H.; Mavumengwana, V.; Li, H.B. Bioactive compounds and biological functions of garlic (allium sativum L.). *Foods,* **2019**, *8*(7), 246.
[http://dx.doi.org/10.3390/foods8070246] [PMID: 31284512]

[99] Ma, H.; Johnson, S.; Liu, W.; DaSilva, N.; Meschwitz, S.; Dain, J.; Seeram, N. Evaluation of polyphenol anthocyanin-enriched extracts of blackberry, black raspberry, blueberry, cranberry, red raspberry, and strawberry for free radical scavenging, reactive carbonyl species trapping, anti-glycation, anti-β-amyloid aggregation, and mic. *Int. J. Mol. Sci.,* **2018**, *19*(2), 461.
[http://dx.doi.org/10.3390/ijms19020461] [PMID: 29401686]

[100] Mohamed, E.A.; Ahmed, H.I.; Zaky, H.S.; Badr, A.M. Sesame oil mitigates memory impairment, oxidative stress, and neurodegeneration in a rat model of Alzheimer's disease. A pivotal role of NF-κB/p38MAPK/BDNF/PPAR-γ pathways. *J. Ethnopharmacol.,* **2021**, *267*, 113468.
[http://dx.doi.org/10.1016/j.jep.2020.113468] [PMID: 33049345]

CHAPTER 3

Nano carriers Containing Bioactive Compounds for Targeting Neurodegenerative Disorders

Abstract: Neurodegenerative Disorders (NDs) are caused by a major loss of neurons both structurally and functionally. The current method of disease management has now encountered several side effects and also the progressive nature of NDs always evokes patients to switch to other drugs. The helpful impact of medicinal plants in these situations has been attributed to their demonstration through several cellular and molecular processes. Natural phytochemicals have served as a good and reliable resource for disease treatment and management. A few neuroprotective mechanisms of these phytochemicals include the reduction in inflammatory responses, the inhibition of pro-inflammatory cytokines' functional aspects, such as tumor growth, and the enhancement of antioxidant qualities. Prevention strategies of these phytoconstituents for NDs heavily rely on variations in transcription and transduction pathways. Aging is one of the main causes of NDs and disease progression, which are mostly brought on by protein loss, oxidative and inflammatory stress, environmental changes, and other factors. Neurodegenerative disorders can be treated with natural substances. Some of the therapeutic herbs for preventing NDs are ginseng, *Withania somnifera, Bacopa monnieri, ginkgo biloba*, and others.

Keywords: Bioactive compounds, *Bacopa monnieri*, Cytokines, Natural phytochemicals, *Withania somnifera*.

INTRODUCTION

Worldwide, NDs impact millions of individuals, and they arise due to a lack of functional mechanisms of nerve cells, which progressively and gradually result in the death of nerve cells of the sympathetic and parasympathetic nervous system [1]. The risk of acquiring a neurological disease as well as its progression increases significantly with aging. The rising average lifespan is also a possible factor that more people will suffer from NDs in the ensuing decades. We need to increase our knowledge of the factors that contribute to neurodegenerative disorders and create fresh strategies for both prevention and therapy [2]. In susceptible people, NDs are characterized by gradual neuronal loss. Important physiological processes such as oxidative stress, proteotoxic stress, neuroinflammation, and apoptosis led to neuronal dysfunction and eventual death

Shivendra Mani Tripathi, Sudhanshu Mishra, Rishabha Malviya & Smriti Ojha

[3]. Because of the subsequent immune-based activation of the central nervous system (CNS), NDs are thought to impose a substantial burden on the population and the healthcare system. Activating or modifying the immunogenic responses can aid in brain regeneration and repair [4]. Nevertheless, immune-mediated conditions like infections, immunological-mediated diseases, and neurodegeneration can also be brought on by immune activation. Over time neurodegeneration slowly but steadily progresses when neurons and axons in the central nervous system begin to die [5]. This causes abnormalities in cellular function and, ultimately, cell death. The following symptoms, which include a variety of deficits including poor memory, lack of coordination, and the total incapacity to carry out daily tasks necessary for leading a healthy lifestyle, appear throughout the degenerative phase. The four most prevalent disorders on the list of NDs are Amyotrophic Lateral Sclerosis (ALS), Parkinson's disease (PD), Alzheimer's disease (AD), and Huntington's disease (HD) [6]. These disorders are highly correlated with rising age, compromised immunity, environmental circumstances, and the affected individual's genetic makeup.

By improving the delivery of drugs to the central nervous system, nano-formulations and nanoparticles such as liposomes, Solid Lipid Nanoparticles (SLNs), polymeric nanoparticles, and magnetic nanoparticles provide innovative approaches to treating neurological disorders. While SLNs offer superior stability and prolonged drug release, liposomes enhance drug stability and release [7].

Magnetic nanoparticles provide for precise, external field-guided targeting, whereas polymeric nanoparticles provide controlled release with targeted delivery. Stability, flexibility, and biocompatibility are still issues despite the various advantages [8].

Challenges to the intranasal delivery of drugs include patient compliance, fast mucociliary clearance, low drug bioavailability, and nasal anatomical variability. To maximize effectiveness, tactics including mucoadhesive agents, penetration enhancers, and sophisticated delivery systems are being developed. Innovative formulations, patient education, and customized technology for better therapy results are needed to overcome these challenges [9].

Natural remedies (Table **1**), particularly Chinese herbal remedies, have gained attention recently as potentially effective treatment options for AD because of their ability to target the disease on several different levels [10]. An essential biological barrier separating the CNS from the circulatory system is the Blood-Brain Barrier (BBB). The BBB can keep many drugs from entering the nervous system in addition to blocking poisons. According to reports, 98% of small-molecule drugs with neuroprotective properties are unable to penetrate this biological barrier [11, 12]. Furthermore, if a phytochemical can permeate the BBB, a low solubility index in brain tissue is another challenge. In most cases, patients require an increased dosage to obtain sufficient drug plasma concentrations. Nevertheless, toxicity at large dosages could result in harm [13]. With the development of nanotechnology in recent years, scientists have discovered that drugs contained in nanoparticles can cross the BBB to reach the affected area, boosting the concentration of drugs in the nervous system and offering promising opportunities for better drug delivery [14]. The latest developments in the application of nanomaterials to the treatment of a variety of NDs and important functions of several natural substances, particularly Chinese herbal medicine-loaded nanoparticles may open up new avenues for the development of dosage forms for the condition in the future [15].

Table 1. Medicinal plants with therapeutic potential for management of neuro-degenerated disorder.

Disease	Medicinal Plant	Phytoconstituents	Bioactivity	Outcomes	Ref
Alzheimer's Disease	*Ginkgo biloba*	Flavonoids, terpenoids	Antioxidant, anti-inflammatory, neuroprotective	Improved memory, cognitive function	[16]
Parkinson's Disease	*Mucuna pruriens*	L-DOPA, flavonoids	Dopaminergic, antioxidant	Improved motor symptoms	[17]
Glaucoma	*Ginkgo biloba, Cannabis sativa (CBD)*	Flavonoids, cannabinoids	Neuroprotective, antioxidant	Reduced intraocular pressure, improved vision	[18]
Multiple Sclerosis	*Panax ginseng*	Ginsenosides	Anti-inflammatory, neuroprotective	Reduced relapse rate, improved symptoms	[19]
Friedreich's Ataxia	*Ginkgo biloba, Bacopa monnieri*	Flavonoids, bacosides	Neuroprotective, antioxidant	Improved motor coordination, quality of life	[20]
Prion Disease	*Curcuma longa (Turmeric)*	Curcuminoids	Antioxidant, anti-inflammatory	Reduced prion accumulation, improved cognitive function	[21]
Amyotrophic Lateral Sclerosis	*Withania somnifera*	Withanolides, alkaloids	Neuroprotective, anti-inflammatory	Improved motor function, quality of life	[22]
Motor Neuron Disease	*Rhodiola rosea*	Salidroside, flavonoids	Neuroprotective, antioxidant	Improved motor function, delayed disease progression	[23]

(Table 1) cont.....

Disease	Medicinal Plant	Phytoconstituents	Bioactivity	Outcomes	Ref
Epilepsy	*Cannabis sativa* (CBD)	Cannabidiol	Anticonvulsant, neuroprotective	Reduced seizure frequency, improved quality of life	[24]
Spinal Muscular Atrophy	*Centella asiatica*	Triterpenoids, flavonoids	Neuroprotective, anti-inflammatory	Improved motor function, quality of life	[25]
Spinocerebellar Ataxia	*Ginkgo biloba, Bacopa monnieri*	Flavonoids, bacosides	Neuroprotective, antioxidant	Improved motor coordination, quality of life	[26]

Barriers to Brain Drug Delivery

Rather than curing the underlying causes of NDs, current therapy has helped to slow the disease's course. The barrier made up of glial cells and capillary walls separating brain cells from plasma, as well as the choroid plexus's barrier separating plasma from cerebrospinal fluid, is referred to as the BBB [27]. Vascular endothelial cells make up its fundamental structure, which is linked by adhesion junction and tight junction proteins. To treat neurological conditions, nanoformulations promote drug delivery across the Blood-Brain Barrier (BBB) by employing processes such as adsorptive-mediated transport, receptor-mediated transcytosis, and transcellular diffusion. Liposomes, Solid Lipid Nanoparticles (SLNs), and polymeric nanoparticles are examples of nanoparticles that may be functionalized with ligands or antibodies to target certain brain receptors, hence enhancing drug absorption [28].

Localized distribution is further aided by different strategies like magnetic nanoparticles and ultrasound-assisted targeting. Through the olfactory and trigeminal nerves, intranasal administration bypasses the blood-brain barrier. These methods provide specific therapy for CNS disorders, improve drug absorption, and reduce adverse effects [29].

To preserve the integrity of the central nervous system's internal environment, the BBB can limit the entry of peripheral macromolecular proteins, cytotoxic chemicals, and peripheral immune cells into the nerve center [30, 31]. The majority of formulations fall short when it comes to the neurodegenerative issue that lies behind the BBB. The effective use of NDs as an intervention is restricted by their incapacity to deliver adequate dosages to the brain. The fact that there are not many effective treatment choices for NDs can be attributed to the mature nature of the BBB and the low penetration potency of most drugs [32]. For brain diseases, only certain phytochemicals are BBB permeable. The permeability of phytochemicals across the BBB is crucial to their current therapeutic applications for brain disorders, yet there is ongoing debate over the appropriate dosage. As was previously mentioned, a drug's prescribed dosage is merely one factor in deciding whether or not it has a high level of therapeutic efficacy [33]. Nevertheless, given the extremely scant data at hand, we endeavor to compile additional guidelines and herewith present instances of phytochemicals that, depending on whether their administered dosage was greater or lower than 10 mg/kg/day, are more or less likely to exhibit BBB permeability. Regardless of the delivery route, this dosage measurement merely serves as a recommendation because all existing routes experience substantial loss during blood circulation or stomach ingestion [34]. The most researched phytochemical subtypes among the known choices are polyphenols and flavonoids which were thought to have neuroprotective properties and to be effective in treating a wide range of diseases, such as cancer, metabolic syndrome, and cardiovascular disorders [35].

BBB acts as a diffusion barrier, keeping chemicals from the blood out of the brain and preserving homeostasis and normal brain function. A tight brain capillary network in the BBB is created by the fusion of several brain cells, including basal membranes, tight junctions, neurons, astrocytes, and brain microvascular endothelial cells [36]. Lipid-based nanocarriers like liposomes and solid lipid nanoparticles, as well as polymer-based nanocarriers like dendrimers and nanoparticles, have been studied for their potential to deliver bioactive compounds for neurodegenerative disorders. These nanocarriers can be modified with targeting ligands to improve their specificity and efficacy [37]. Compounds with certain physicochemical characteristics, such as molecular size and weight, surface electrical properties, lipophilicity, and surface charge, can pass through the BBB. Passive diffusion allows certain tiny chemicals (including carbon dioxide, oxygen, ethanol, and barbiturates) to effortlessly move through the BBB, certain macromolecular materials, however, are impervious to it [38]. Hydrophilic molecules are also transported by receptor-mediated transport systems. Moreover, it is known that certain pathological conditions can cause the blood-brain barrier to become less taut, which permits drugs to seep into the brain. Ultimately, the transportation of cargo across the BBB can be improved by the introduction of specialized drug carriers like nanoparticles as shown in Fig. (**1**). We go over a couple of these examples below [39]. The presence of the BBB, which prevents many drugs from entering the brain in appropriate concentrations, is a significant issue when treating NDs. Many targeted tactics are being studied to deliver various therapeutic agents *via* the BBB, even though the BBB prevents many treatments that can be used to treat NDs from entering the brain. Among the most promising methods are nanotechnology and the development of nanocarriers [40, 41]. Pharmacokinetic properties have a major role in determining the effectiveness of therapeutic agents that are administered systematically. It is a difficult voyage that, for the most part, does not favor the therapeutic molecules from the point of administration to the target site (the brain in this example) [42]. The presence of different embedded plasma proteins is monitored as certain drugs have a strong binding to these proteins, which reduces the amount of the therapeutic ingredient in circulation and, in turn, the amount of free active pharmaceutical ingredients that can reach the brain [43]. Furthermore, only a small percentage of therapeutic moiety remains in the bloodstream after being removed at a high rate by the primary clearance organs. Additionally, the amount of drug absorption is limited by the way the drug interacts with the target cells. More precisely, pharmacological compounds can have an impact on cells that results in modifications to cell shape, membrane potential, or even channel blockage. This fleeting action may restrict how the cell responds to the drug molecule being delivered and how much of it is absorbed. Small lipophilic pharmacological compounds are often appropriate for brain administration [44, 45].

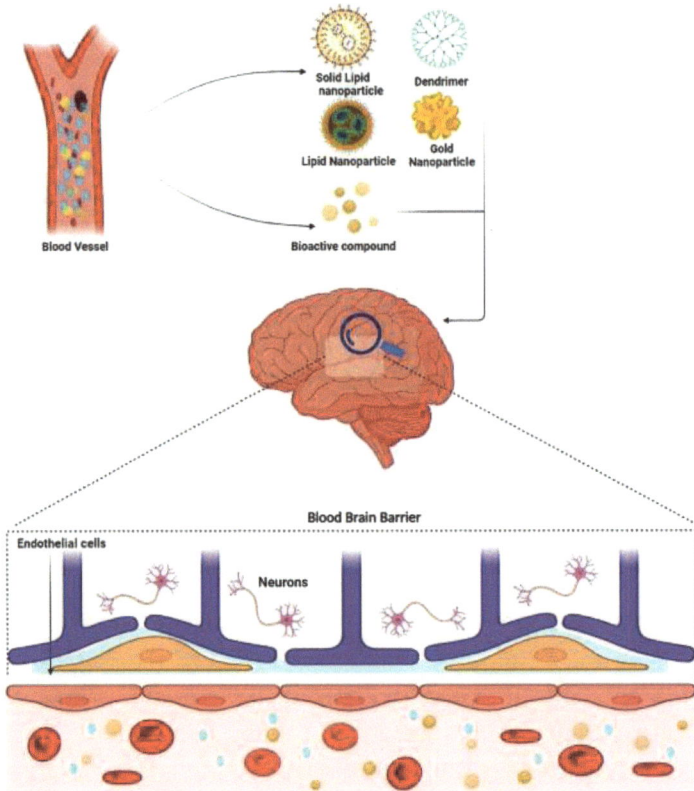

Fig. (1). Nanocarriers containing bioactive compounds for targeting neurodegenerative disorders.

Potential Role of Herbal Nanotherapeutics for the Management of Neurodegenerative Disorders

The above-mentioned drawbacks of the existing medicines combined with BBB limitations have created a requirement for the development of novel drug delivery techniques for the management of NDs. Among the methods used, targeted delivery of therapeutic agents to the central nervous system has shown promise and safety with nanotechnology [46]. This technology works with materials at the nanoscale, typically between 1 and 1000 nm, and possesses molecular interaction ability with various bio-molecules. Synthetic polymers, inorganic elements, and a few natural polymers have all been used to create nanoparticles [47]. Due to their high drug-loading capacity, low systemic toxicity, increased drug permeability and transport to targeted tissues, and improved drug stability, nanocarriers are a viable platform for the management and treatment of NDs [48].

Nevertheless, the size, type, polarity, and surface chemistry of the nanocarriers determine how well they penetrate the BBB. Furthermore, the polysorbate surface coating can aid in avoiding transmembrane efflux mechanisms. Structural modifications while developing various nanocarriers with ligands and cell-penetrating peptides make them suitable candidates for targeted brain administration [49]. To solve several concerns concerning the therapeutic potential of phytochemicals, such as solubility and tissue accumulation, drug stability, specificity, and passage through the BBB, nanoparticles may prove to be a powerful delivery system [50]. Materials or structures with a nanoscale scale are referred to as nanoparticles. Because of their size, the nanoparticles may be able to get through cell barriers. Various nanoparticle formulas have advantages and disadvantages of their own when used as drug delivery carriers; nevertheless, the characterization and properties that can be altered as needed are what allow for the optimization of phytochemicals against NDs [51]. Furthermore, the ability of nanoparticles to distribute and release drugs specifically may significantly reduce the dosage of medication taken orally or intravenously. The inherent qualities of the intended medication formulation, such as electrical charges or water solubility, are typically used in loading its cargo [52]. A successful nanomedicine also needs to be released at the site of administration. Conditional stimulus-responsive mechanisms are used to ensure effective regulated release. Redox-potential-based stimulus, pH-based stimulus, temperature-based stimulus, and other external stimuli like laser, magnetic, or ultrasonic waves are a few notable examples [53]. Researchers have discovered various techniques for using nanoparticles to facilitate phytochemical brain treatments. First, phyto-bioactive chemicals that have been shown to have the ability to influence oxidative stress and inflammation—two factors known to be significant in brain-associated degenerative conditions—are loaded into nano-delivery devices [54]. Many compelling studies conducted in the last few years have demonstrated the great potential of phytochemical-loaded nanocarriers to combat diseases of the brain and central nervous system, such as glioblastoma, rheumatoid arthritis, and neurodegenerative disorders. Meanwhile, the brain can access another channel across the BBB by altering it through transporter uptake or other mechanisms [55]. The solute carrier transporters and the ATP-binding cassette transporters, which are expressed at the luminal side of the BBB and actively efflux from the endothelium into the blood, are key players in maintaining drug concentration and distribution phytochemicals. It is possible to use phytochemicals, particularly those with smaller molecular sizes, to decorate the nanocarriers for transcytosis, a transporter-mediated transcytosis [56].

Researchers are also investigating alternative translocating mechanisms that utilize the distinct action of the BBB upon contact with phytochemicals. Menthol, for example, may facilitate the translocation of nanoparticles across the BBB, which may be attributed to a notable increase in cell membrane fluidity and a subsequent decrease in membrane potential, facilitating drug transport [57]. By encouraging internalization, disturbing tight-junction integrity, and BBB rupture, the functionalized menthol nanocarriers allowed nanotherapeutics to pass through BBB. Sinapic acid and mustard act as a novel BBB-permeable ligand for effective medication administration into the brain. Its mediating transporter(s) across the BBB are still very much up in the air [58]. Additionally, phytochemicals themselves may be functionalized into a portion of the nanoparticle to aid in their absorption *via* the blood-brain barrier and, once released into the brain microenvironment, they employ the phytochemical's therapeutic properties for medical intervention [59].

In two-year research, Bushen Capsules (BSC) improved or maintained cognitive function in people with amnestic Moderate Cognitive Impairment (aMCI) when compared to a placebo. According to functional MRI research, BSC therapy provides long-term cognitive improvements by controlling aberrant brain activity. This study shows the viability of employing neuroimaging biomarkers in clinical trials and emphasizes the potential of Traditional Chinese Medicine (TCM) for early Alzheimer's treatment [60].

For intranasal administration in Alzheimer's disease, a study developed and analyzed two CUR formulations: CUR/HP-β-CD inclusion complexes and CUR-encapsulated chitosan-coated poly(lactic-co-glycolic acid) nanoparticles (CUR-CS-PLGA-NPs). In comparison to CUR-CS-PLGA-NPs, the CUR/HP-β-CD inclusion complexes demonstrated improved pharmacokinetics, higher cellular absorption, and superior stability, with significantly higher plasma and brain CUR levels. Although both formulations demonstrated anti-inflammatory and antioxidant properties, CUR/HP-β-CD outperformed the other, making it a better option for CUR administration in the treatment of Alzheimer's disease [61].

Berberine-loaded Nanostructured Lipid Carriers (Berb-NLCs) were developed to improve the poor absorption and brain penetration of berberine for Alzheimer's disease treatment. The optimized formulation (Trial-5) showed good particle size, entrapment efficiency, and sustained release. *In vivo* studies demonstrated improved behavioral parameters in Alzheimer's model rats, indicating the potential of Berb-NLCs for effective brain targeting and therapeutic efficacy [62].

Levodopa (L-DOPA)-loaded sub-50 nm nanoparticles were developed using tannic acid/polyvinyl alcohol to deliver L-DOPA through the brain-lymphatic vasculature, bypassing peripheral circulation. In a Parkinson's disease rat model, the formulation significantly improved dopaminergic neuron function, reduced movement disorders, and alleviated oxidative stress, demonstrating high biocompatibility and potential for clinical treatment of Parkinson's disease [63].

Case Studies of Successful Nanocarrier Applications

Berberine-Loaded Nanostructured Lipid Carriers (NLCs)

Designed specifically for Alzheimer's disease, berberine-loaded NLCs aim to overcome the compound's poor absorption and limited brain penetration. Studies have shown that these NLCs significantly increase berberine's bioavailability in the brain. In animal models of Alzheimer's disease, treatment with these NLCs resulted in enhanced cognitive performance and reduced oxidative stress markers. The encapsulation technology not only stabilized berberine but also facilitated sustained release, reducing the frequency of administration and minimizing potential side effects [64].

Levodopa-Encapsulated Nanoparticles

Parkinson's disease treatment has long relied on levodopa, but its systemic side effects and low brain bioavailability pose challenges. Levodopa-encapsulated nanoparticles, developed using tannic acid and polyvinyl alcohol, have demonstrated a unique capability to bypass peripheral circulation through brain-lymphatic vasculature targeting. Animal studies reported significant improvements in dopaminergic neuron function, reduced oxidative damage, and alleviation of motor symptoms. These nanoparticles also showed a prolonged therapeutic effect, maintaining motor improvements for up to one-week post-treatment [65].

Curcumin-Loaded Chitosan-Coated Nanoparticles

Curcumin, a potent anti-inflammatory and antioxidant compound, faces challenges in CNS delivery due to its poor water solubility and rapid metabolism. Chitosan-coated nanoparticles have emerged as a promising solution, enhancing curcumin's stability, cellular uptake, and bioavailability. Comparative analyses of curcumin-loaded nanoparticles with traditional formulations demonstrated superior therapeutic outcomes, including reduced neuroinflammation and improved cognitive function in Alzheimer's disease models [66].

Multifunctional Nanoparticles for Synergistic Therapy

Recent advances also include dual-loaded nanoparticles carrying both levodopa and curcumin for Parkinson's disease. These multifunctional systems leverage the synergistic effects of both compounds, combining the neuroprotective properties of curcumin with the dopamine-restorative action of levodopa. In animal studies, these nanoparticles not only improved motor functions but also showed reduced systemic toxicity due to their targeted delivery mechanism [67]. These examples underscore the potential of nanocarriers to enhance the efficacy of bioactive compounds, particularly in neurodegenerative disorders.

Future Directions and Practical Insight

Emerging technologies, such as AI-driven drug design and advanced imaging techniques, can further optimize nanocarrier development. AI can predict nanocarrier interactions with the BBB, while imaging techniques can assess their biodistribution in real time. Collaborative efforts between academia, industry, and regulatory bodies are vital to accelerate the clinical translation of these technologies. Additionally, patient-centric approaches, including the customization of formulations based on genetic profiles, can pave the way for precision medicine in neurodegenerative disorders. By addressing production challenges, ensuring safety through rigorous testing, and leveraging cutting-edge technologies, nanocarriers have the potential to transform the landscape of neurodegenerative disorder management.

Challenges

The main hallmark of NDs, such as Alzheimer's disease and Parkinson's, is neuronal death. Consequently, the most anticipated course of treatment for these conditions is neurogenesis. Nevertheless, there are still several important characteristics that make medication transport to the brain difficult, such as the BBB, lipophilicity, the drug's molecular weight, *etc.* These elements restrict the therapeutic potency of drugs and increase the difficulty of treating NDs.

Therefore, targeted medication delivery to the brain *via* nanoparticles has been investigated recently for neurogenesis, and it offers a viable platform for enhancing therapeutic approaches. Notwithstanding these possible benefits, there are many difficult issues of nanocarrier-mediated medication administration, such as production, safety, and legislation. The main determinants of nanoparticle toxicity are their size, shape, ionic dissolutions, and surface charge. According to the official guidelines on nanotoxicity, these characteristics should be taken into account when creating drug-delivery systems based on nanoparticles. Furthermore, a thorough analysis of these nanocarriers' approval, taking into account their effects on the environment and human health, is necessary. A large body of research recommends several changes to reduce the toxicity related to the size and charge, such as adding biodegradable or intrinsic biomolecules to the surface. Regarding nanocarrier production, batch-to-batch consistency in terms of size and content should be maintained [68]. The intranasal administration of nano-DOPA, a PLGA-based L-DOPA nanoparticle, significantly improved motor function in a rat Parkinson's disease model, with prolonged effects lasting 24 hours and even one week after treatment discontinuation. Challenges include overcoming low drug solubility, ensuring consistent dosing, and achieving optimal brain delivery for chronic use in humans [69]. A study presents a novel therapy for Parkinson's disease using biodegradable nanoparticles, consisting of NH2–PEO–PCL. These nanoparticles transport L-DOPA and curcumin simultaneously, providing a synergistic therapeutic effect. The nanoparticles are blood-compatible, low cytotoxic, and could be applied with lower concentrations, new routes, and higher treatment tolerance [70].

Despite their promise, nanocarriers face significant challenges, including production scalability, safety concerns, and regulatory hurdles. Batch-to-batch consistency in particle size and drug loading remains a production challenge. For example, high-pressure homogenization a common method for nanoparticle synthesis requires stringent controls to maintain uniformity [71]. Practical solutions include adopting advanced manufacturing techniques like microfluidics, which ensures precise size control and reproducibility. Safety concerns primarily arise from the potential toxicity of nanomaterials, which is influenced by factors like particle size, surface charge, and biodegradability [72]. Cationic nanoparticles, while efficient in crossing the BBB, may cause cytotoxicity. Incorporating biodegradable polymers such as poly(lactic-co-glycolic acid) (PLGA)mitigates these risks. Thorough *in vitro* and *in vivo* toxicological studies are essential to evaluate long-term safety. This would support the scientific community in expanding or identifying useful nanoparticles for medicinal or diagnostic uses [73].

CONCLUSION

The special characteristics of the BBB and how it prevents the application of phytochemicals to brain disorders are highlights of this chapter. Despite their low BBB penetration, most phytochemicals have been shown to have a promising role in brain therapy. To maximize the usage of phytochemicals—both permeable and non-permeable—across the blood-brain barrier, a potential nano-delivery platform was proposed. This could open up new possibilities for the clinical application of phytochemicals in brain disorders. Many of the active pharmacological medicines available for the treatment of NDs are unable to penetrate the blood-brain barrier and yield impressive outcomes in clinical studies. An emerging field called nanomedicine holds the promise of removing the challenges that come with standard medicine. Brain targeting technologies based on nano biomaterials seem to be the alternatives being explored for resolving these problems since they allow the change of particular neural pathways in designated brain areas and target those specific brain regions to achieve the intended therapeutic effects. In a similar vein, the combination of gene therapy and nanotechnology may improve gene therapy's efficacy in treating non-DNA disorders. Recent advances in nano biomaterials-oriented techniques have improved the effectiveness of existing neural stem cell (NSC) differentiation-based therapies. It also aided in comprehending the molecular underpinnings of the relevant process. One of the safest and most effective methods for NDs may be the creation of NSCs-targeting technologies. It is possible to draw the conclusion that the various treatment techniques used to treat NDs include the use of nanocarriers, phytoconstituents, gene therapy, and NSCs.

REFERENCES

[1]   Tindle, J.; Tadi, P. *Neuroanatomy, Parasympathetic Nervous System*; StatPearls Publishing, **2020**.

[2]   Wareham, L.K.; Liddelow, S.A.; Temple, S.; Benowitz, L.I.; Di Polo, A.; Wellington, C.; Goldberg, J.L.; He, Z.; Duan, X.; Bu, G.; Davis, A.A.; Shekhar, K.; Torre, A.L.; Chan, D.C.; Canto-Soler, M.V.; Flanagan, J.G.; Subramanian, P.; Rossi, S.; Brunner, T.; Bovenkamp, D.E.; Calkins, D.J. Solving neurodegeneration: common mechanisms and strategies for new treatments. *Mol. Neurodegener.,* **2022**, *17*(1), 23.
[http://dx.doi.org/10.1186/s13024-022-00524-0] [PMID: 35313950]

[3]   Nayab, D.E.; Din, F.; Ali, H.; Kausar, W.A.; Urooj, S.; Zafar, M.; Khan, I.; Shabbir, K.; Khan, G.M. Nano biomaterials based strategies for enhanced brain targeting in the treatment of neurodegenerative diseases: an up-to-date perspective. *J. Nanobiotechnology,* **2023**, *21*(1), 477.
[http://dx.doi.org/10.1186/s12951-023-02250-1] [PMID: 38087359]

[4]   Xu, J.; Ma, C.; Hua, M.; Li, J.; Xiang, Z.; Wu, J. CNS and CNS diseases in relation to their immune system. *Front. Immunol.,* **2022**, *13*, 1063928.
[http://dx.doi.org/10.3389/fimmu.2022.1063928] [PMID: 36466889]

[5]   Passaro, A.P.; Lebos, A.L.; Yao, Y.; Stice, S.L. Immune Response in Neurological Pathology: Emerging Role of Central and Peripheral Immune Crosstalk. *Front. Immunol.,* **2021**, *12*, 676621.
[http://dx.doi.org/10.3389/fimmu.2021.676621] [PMID: 34177918]

[6]   Garofalo, M.; Pandini, C.; Bordoni, M.; Pansarasa, O.; Rey, F.; Costa, A.; Minafra, B.; Diamanti, L.; Zucca, S.; Carelli, S.; Cereda, C.; Gagliardi, S. Alzheimer's, parkinson's disease and amyotrophic lateral sclerosis gene expression patterns divergence reveals different grade of RNA metabolism involvement. *Int. J. Mol. Sci.,* **2020**, *21*(24), 9500.
[http://dx.doi.org/10.3390/ijms21249500] [PMID: 33327559]

[7]   Mishra, S.; Shah, H.; Patel, A.; Tripathi, S.M.; Malviya, R.; Prajapati, B.G. Applications of Bioengineered Polymer in the Field of Nano-Based Drug Delivery. *ACS Omega,* **2024**, *9*(1), 81-96.
[http://dx.doi.org/10.1021/acsomega.3c07356] [PMID: 38222544]

[8]   Shukla, S.; Tiwari, S.; Bhattacharya, R.; Ojha, S.; Mishra, S.; Gupta, S.K. Nanotechnology-Based Approaches for Nose-to-Brain Drug Delivery in Neurodegenerative Diseases. *Lett. Drug Des. Discov.,* **2024**, *21*(11), 1913-1921.
[http://dx.doi.org/10.2174/1570180820666230622120759]

[9]   Ojha, S.; Mishra, S. Nanostructured Lipid Carriers for Targeting Central Nervous System: Recent Advancements. *Micro Nanosyst.,* **2023**, *15*(2), 82-91.
[http://dx.doi.org/10.2174/1876402915666230518121949]

[10]   Liu, N.; Ruan, J.; Li, H.; Fu, J. Nanoparticles loaded with natural medicines for the treatment of Alzheimer's disease. *Front. Neurosci.,* **2023**, *17*, 1112435.
[http://dx.doi.org/10.3389/fnins.2023.1112435] [PMID: 37877008]

[11]   Teleanu, R.I.; Preda, M.D.; Niculescu, A.G.; Vladâcenco, O.; Radu, C.I.; Grumezescu, A.M.; Teleanu, D.M. Current Strategies to Enhance Delivery of Drugs across the Blood–Brain Barrier. *Pharmaceutics,* **2022**, *14*(5), 987.
[http://dx.doi.org/10.3390/pharmaceutics14050987] [PMID: 35631573]

[12]   Kadry, H.; Noorani, B.; Cucullo, L. A blood–brain barrier overview on structure, function, impairment, and biomarkers of integrity. *Fluids Barriers CNS,* **2020**, *17*(1), 69.
[http://dx.doi.org/10.1186/s12987-020-00230-3] [PMID: 33208141]

[13]   Liu, Y.; Chen, Z.; Li, A.; Liu, R.; Yang, H.; Xia, X. The Phytochemical Potential for Brain Disease Therapy and the Possible Nanodelivery Solutions for Brain Access. *Front. Oncol.,* **2022**, *12*, 936054.
[http://dx.doi.org/10.3389/fonc.2022.936054] [PMID: 35814371]

[14]   Mittal, K.R.; Pharasi, N.; Sarna, B.; Singh, M.; Rachana, ; Haider, S.; Singh, S.K.; Dua, K.; Jha, S.K.; Dey, A.; Ojha, S.; Mani, S.; Jha, N.K. Nanotechnology-based drug delivery for the treatment of CNS disorders. *Transl. Neurosci.,* **2022**, *13*(1), 527-546.
[http://dx.doi.org/10.1515/tnsci-2022-0258] [PMID: 36741545]

[15]   Waris, A.; Ali, A.; Khan, A.U.; Asim, M.; Zamel, D.; Fatima, K.; Raziq, A.; Khan, M.A.; Akbar, N.; Baset, A.; Abourehab, M.A.S. Applications of Various Types of Nanomaterials for the Treatment of Neurological Disorders. *Nanomaterials (Basel),* **2022**, *12*(13), 2140.
[http://dx.doi.org/10.3390/nano12132140] [PMID: 35807977]

[16]   Singh, S.K.; Srivastav, S.; Castellani, R.J.; Plascencia-Villa, G.; Perry, G. Neuroprotective and Antioxidant Effect of Ginkgo biloba Extract Against AD and Other Neurological Disorders. *Neurotherapeutics,* **2019**, *16*(3), 666-674.
[http://dx.doi.org/10.1007/s13311-019-00767-8] [PMID: 31376068]

[17]   Khazdair, M.R.; Kianmehr, M.; Anaeigoudari, A. Effects of medicinal plants and flavonoids on Parkinson's disease: A review on basic and clinical evidences. *Adv. Pharm. Bull.,* **2020**, *11*(2), 224-232.
[http://dx.doi.org/10.34172/apb.2021.026] [PMID: 33880344]

[18]   Ige, M.; Liu, J. Herbal medicines in glaucoma treatment. *Yale J. Biol. Med.,* **2020**, *93*(2), 347-353.
[PMID: 32607093]

[19]   Cho, I.H. Effects of Panax ginseng in neurodegenerative diseases. *J. Ginseng Res.,* **2012**, *36*(4), 342-353.

[http://dx.doi.org/10.5142/jgr.2012.36.4.342] [PMID: 23717136]

[20]   Nathan, P.J.; Tanner, S.; Lloyd, J.; Harrison, B.; Curran, L.; Oliver, C.; Stough, C. Effects of a combined extract of *Ginkgo biloba* and *Bacopa monniera* on cognitive function in healthy humans. *Hum. Psychopharmacol.,* **2004**, *19*(2), 91-96.
[http://dx.doi.org/10.1002/hup.544] [PMID: 14994318]

[21]   Górka, M.; Białoń, N.; Bieczek, D.; Górka, D. [Neuroprotective effect of curcumin and its potential use in the treatment of neurodegenerative diseases]. *Postepy Biochem.,* **2023**, *69*(1), 18-25.
[http://dx.doi.org/10.18388/pb.2021_472] [PMID: 37493562]

[22]   Dutta, K.; Patel, P.; Julien, J.P. Protective effects of Withania somnifera extract in SOD1^{G93A} mouse model of amyotrophic lateral sclerosis. *Exp. Neurol.,* **2018**, *309*, 193-204.
[http://dx.doi.org/10.1016/j.expneurol.2018.08.008] [PMID: 30134145]

[23]   Shen, J.; Chen, S.; Li, X.; Wu, L.; Mao, X.; Jiang, J.; Zhu, D. Salidroside Mediated the Nrf2/GPX4 Pathway to Attenuates Ferroptosis in Parkinson's Disease. *Neurochem. Res.,* **2024**, *49*(5), 1291-1305.
[http://dx.doi.org/10.1007/s11064-024-04116-w] [PMID: 38424396]

[24]   Lazarini-Lopes, W.; Do Val-da Silva, R.A.; da Silva-Júnior, R.M.P.; Leite, J.P.; Garcia-Cairasco, N. The anticonvulsant effects of cannabidiol in experimental models of epileptic seizures: From behavior and mechanisms to clinical insights. *Neurosci. Biobehav. Rev.,* **2020**, *111*, 166-182.
[http://dx.doi.org/10.1016/j.neubiorev.2020.01.014] [PMID: 31954723]

[25]   Shinomol, G.K.; Muralidhara, ; Bharath, M.M. Exploring the Role of "Brahmi" (Bacopa monnieri and Centella asiatica) in Brain Function and Therapy. *Recent Pat. Endocr. Metab. Immune Drug Discov.,* **2011**, *5*(1), 33-49.
[http://dx.doi.org/10.2174/187221411794351833] [PMID: 22074576]

[26]   Wu, C-H.; Huang, D-S.; Lin, H-Y.; Lee-Chen, G-J.; Hsieh, H-M.; Lin, J.Y. Treatment with a Ginkgo biloba extract, EGb 761, inhibits excitotoxicity in an animal model of spinocerebellar ataxia type 17. *Drug Des. Devel. Ther.,* **2016**, *10*, 723-731.
[http://dx.doi.org/10.2147/DDDT.S98156] [PMID: 26937174]

[27]   Solár, P.; Zamani, A.; Kubíčková, L.; Dubový, P.; Joukal, M. Choroid plexus and the blood–cerebrospinal fluid barrier in disease. *Fluids Barriers CNS,* **2020**, *17*(1), 35.
[http://dx.doi.org/10.1186/s12987-020-00196-2] [PMID: 32375819]

[28]   Pinheiro, R.G.R.; Coutinho, A.J.; Pinheiro, M.; Neves, A.R. Nanoparticles for targeted brain drug delivery: What do we know? *Int. J. Mol. Sci.,* **2021**, *22*(21), 11654.
[http://dx.doi.org/10.3390/ijms222111654] [PMID: 34769082]

[29]   Asimakidou, E.; Tan, J.K.S.; Zeng, J.; Lo, C.H. Blood–Brain Barrier-Targeting Nanoparticles: Biomaterial Properties and Biomedical Applications in Translational Neuroscience. *Pharmaceuticals (Basel),* **2024**, *17*(5), 612.
[http://dx.doi.org/10.3390/ph17050612] [PMID: 38794182]

[30]   Bors, L.A.; Erdő, F. Overcoming the blood-brain barrier. Challenges and tricks for CNS drug delivery. *Sci. Pharm.,* **2019**, *87*(1), 6.
[http://dx.doi.org/10.3390/scipharm87010006]

[31]   Duong, C.N.; Vestweber, D. Mechanisms Ensuring Endothelial Junction Integrity Beyond VE-Cadherin. *Front. Physiol.,* **2020**, *11*, 519.
[http://dx.doi.org/10.3389/fphys.2020.00519] [PMID: 32670077]

[32]   Lamptey, R.N.L.; Chaulagain, B.; Trivedi, R.; Gothwal, A.; Layek, B.; Singh, J. A Review of the Common Neurodegenerative Disorders: Current Therapeutic Approaches and the Potential Role of Nanotherapeutics. *Int. J. Mol. Sci.,* **2022**, *23*(3), 1851.
[http://dx.doi.org/10.3390/ijms23031851] [PMID: 35163773]

[33]   Sánchez-Martínez, J.D.; Valdés, A.; Gallego, R.; Suárez-Montenegro, Z.J.; Alarcón, M.; Ibañez, E.; Alvarez-Rivera, G.; Cifuentes, A. Blood–Brain Barrier Permeability Study of Potential

Neuroprotective Compounds Recovered From Plants and Agri-Food by-Products. *Front. Nutr.,* **2022**, *9*, 924596.
[http://dx.doi.org/10.3389/fnut.2022.924596] [PMID: 35782945]

[34]   Ren, Y.; Wu, W.; Zhang, X. The feasibility of oral targeted drug delivery: Gut immune to particulates? *Acta Pharm. Sin. B,* **2023**, *13*(6), 2544-2558.
[http://dx.doi.org/10.1016/j.apsb.2022.10.020] [PMID: 37425061]

[35]   Ullah, A.; Munir, S.; Badshah, S.L.; Khan, N.; Ghani, L.; Poulson, B.G.; Emwas, A.H.; Jaremko, M. Important flavonoids and their role as a therapeutic agent. *Molecules,* **2020**, *25*(22), 5243.
[http://dx.doi.org/10.3390/molecules25225243] [PMID: 33187049]

[36]   Alahmari, A. Blood-Brain Barrier Overview: Structural and Functional Correlation. *Neural Plast.,* **2021**, *2021*, 1-10.
[http://dx.doi.org/10.1155/2021/6564585] [PMID: 34912450]

[37]   Lu, H.; Zhang, S.; Wang, J.; Chen, Q. A Review on Polymer and Lipid-Based Nanocarriers and Its Application to Nano-Pharmaceutical and Food-Based Systems. *Front. Nutr.,* **2021**, *8*, 783831.
[http://dx.doi.org/10.3389/fnut.2021.783831] [PMID: 34926557]

[38]   Dong, X. Current strategies for brain drug delivery. *Theranostics,* **2018**, *8*(6), 1481-1493.
[http://dx.doi.org/10.7150/thno.21254] [PMID: 29556336]

[39]   Pawar, B.; Vasdev, N.; Gupta, T.; Mhatre, M.; More, A.; Anup, N.; Tekade, R.K. Current Update on Transcellular Brain Drug Delivery. *Pharmaceutics,* **2022**, *14*(12), 2719.
[http://dx.doi.org/10.3390/pharmaceutics14122719] [PMID: 36559214]

[40]   Barnabas, W. Drug targeting strategies into the brain for treating neurological diseases. *J. Neurosci. Methods,* **2019**, *311*, 133-146.
[http://dx.doi.org/10.1016/j.jneumeth.2018.10.015] [PMID: 30336221]

[41]   Hu, Y.; Gaillard, P.J.; Rip, J.; Hammarlund-Udenaes, M. Blood-to-Brain Drug Delivery Using Nanocarriers. *AAPS Advances in the Pharmaceutical Sciences Series,* **2022**, *33*, 501-526.
[http://dx.doi.org/10.1007/978-3-030-88773-5_16]

[42]   Glassman, P.M.; Muzykantov, V.R. Pharmacokinetic and pharmacodynamic properties of drug delivery systems. *J. Pharmacol. Exp. Ther.,* **2019**, *370*(3), 570-580.
[http://dx.doi.org/10.1124/jpet.119.257113] [PMID: 30837281]

[43]   Charlier, B.; Coglianese, A.; De Rosa, F.; de Grazia, U.; Operto, F.F.; Coppola, G.; Filippelli, A.; Dal Piaz, F.; Izzo, V. The effect of plasma protein binding on the therapeutic monitoring of antiseizure medications. *Pharmaceutics,* **2021**, *13*(8), 1208.
[http://dx.doi.org/10.3390/pharmaceutics13081208] [PMID: 34452168]

[44]   Zhang, R.; Qin, X.; Kong, F.; Chen, P.; Pan, G. Improving cellular uptake of therapeutic entities through interaction with components of cell membrane. *Drug Deliv.,* **2019**, *26*(1), 328-342.
[http://dx.doi.org/10.1080/10717544.2019.1582730] [PMID: 30905189]

[45]   Augustine, R.; Hasan, A.; Primavera, R.; Wilson, R.J.; Thakor, A.S.; Kevadiya, B.D. Cellular uptake and retention of nanoparticles: Insights on particle properties and interaction with cellular components. *Mater. Today Commun.,* **2020**, *25*, 101692.
[http://dx.doi.org/10.1016/j.mtcomm.2020.101692]

[46]   Nguyen, T.T.; Dung Nguyen, T.T.; Vo, T.K.; Tran, N.M.A.; Nguyen, M.K.; Van Vo, T.; Van Vo, G. Nanotechnology-based drug delivery for central nervous system disorders. *Biomed. Pharmacother.,* **2021**, *143*, 112117.
[http://dx.doi.org/10.1016/j.biopha.2021.112117] [PMID: 34479020]

[47]   Khan, I.; Saeed, K.; Khan, I. Nanoparticles: Properties, applications and toxicities. *Arab. J. Chem.,* **2019**, *12*(7), 908-931.
[http://dx.doi.org/10.1016/j.arabjc.2017.05.011]

[48]   Din, F.; Aman, W.; Ullah, I.; Qureshi, O.S.; Mustapha, O.; Shafique, S.; Zeb, A. Effective use of

nanocarriers as drug delivery systems for the treatment of selected tumors. *Int. J. Nanomedicine,* **2017,** *12,* 7291-7309.
[http://dx.doi.org/10.2147/IJN.S146315] [PMID: 29042776]

[49]   Ahlawat, J.; Guillama Barroso, G.; Masoudi Asil, S.; Alvarado, M.; Armendariz, I.; Bernal, J.; Carabaza, X.; Chavez, S.; Cruz, P.; Escalante, V.; Estorga, S.; Fernandez, D.; Lozano, C.; Marrufo, M.; Ahmad, N.; Negrete, S.; Olvera, K.; Parada, X.; Portillo, B.; Ramirez, A.; Ramos, R.; Rodriguez, V.; Rojas, P.; Romero, J.; Suarez, D.; Urueta, G.; Viel, S.; Narayan, M. Nanocarriers as Potential Drug Delivery Candidates for Overcoming the Blood–Brain Barrier: Challenges and Possibilities. *ACS Omega,* **2020,** *5*(22), 12583-12595.
[http://dx.doi.org/10.1021/acsomega.0c01592] [PMID: 32548442]

[50]   Mule, S.; Khairnar, P.; Shukla, R. Recent Advances in Nanocrystals Heralding Greater Potential in Brain Delivery. *Part. Part. Syst. Charact.,* **2022,** *39*(9), 2200087.
[http://dx.doi.org/10.1002/ppsc.202200087]

[51]   Harish, V.; Tewari, D.; Gaur, M.; Yadav, A.B.; Swaroop, S.; Bechelany, M.; Barhoum, A. Review on Nanoparticles and Nanostructured Materials: Bioimaging, Biosensing, Drug Delivery, Tissue Engineering, Antimicrobial, and Agro-Food Applications. *Nanomaterials (Basel),* **2022,** *12*(3), 457.
[http://dx.doi.org/10.3390/nano12030457] [PMID: 35159802]

[52]   Hsu, C.Y.; Rheima, A.M.; Kadhim, M.M.; Ahmed, N.N.; Mohammed, S.H.; Abbas, F.H.; Abed, Z.T.; Mahdi, Z.M.; Abbas, Z.S.; Hachim, S.K.; Ali, F.K.; Mahmoud, Z.H.; Kianfar, E. An overview of nanoparticles in drug delivery: Properties and applications. *S. Afr. J. Chem. Eng.,* **2023,** *46,* 233-270.
[http://dx.doi.org/10.1016/j.sajce.2023.08.009]

[53]   Liu, G.; Lovell, J.F.; Zhang, L.; Zhang, Y. Stimulus-responsive nanomedicines for disease diagnosis and treatment. *Int. J. Mol. Sci.,* **2020,** *21*(17), 6380.
[http://dx.doi.org/10.3390/ijms21176380] [PMID: 32887466]

[54]   Bhattacharya, T.; Soares, G.A.B.; Chopra, H.; Rahman, M.M.; Hasan, Z.; Swain, S.S.; Cavalu, S. Applications of Phyto-Nanotechnology for the Treatment of Neurodegenerative Disorders. *Materials (Basel),* **2022,** *15*(3), 804.
[http://dx.doi.org/10.3390/ma15030804] [PMID: 35160749]

[55]   Fakhri, S.; Abdian, S.; Zarneshan, S.N.; Moradi, S.Z.; Farzaei, M.H.; Abdollahi, M. Nanoparticles in Combating Neuronal Dysregulated Signaling Pathways: Recent Approaches to the Nanoformulations of Phytochemicals and Synthetic Drugs Against Neurodegenerative Diseases. *Int. J. Nanomedicine,* **2022,** *17,* 299-331.
[http://dx.doi.org/10.2147/IJN.S347187] [PMID: 35095273]

[56]   Gameiro, M.; Silva, R.; Rocha-Pereira, C.; Carmo, H.; Carvalho, F.; Bastos, M.; Remião, F. Cellular models and *in vitro* assays for the screening of modulators of P-gp, MRP1 and BCRP. *Molecules,* **2017,** *22*(4), 600.
[http://dx.doi.org/10.3390/molecules22040600] [PMID: 28397762]

[57]   Mhaske, A.; Shukla, S.; Ahirwar, K.; Singh, K.K.; Shukla, R. Receptor-Assisted Nanotherapeutics for Overcoming the Blood–Brain Barrier. *Mol. Neurobiol.,* **2024,** *61*(11), 8702-8738.
[http://dx.doi.org/10.1007/s12035-024-04015-9] [PMID: 38558360]

[58]   Wang, N.; Sun, P.; Lv, M.; Tong, G.; Jin, X.; Zhu, X. Mustard-inspired delivery shuttle for enhanced blood–brain barrier penetration and effective drug delivery in glioma therapy. *Biomater. Sci.,* **2017,** *5*(5), 1041-1050.
[http://dx.doi.org/10.1039/C7BM00133A] [PMID: 28378865]

[59]   Mehan, S.; Arora, N.; Bhalla, S.; Khan, A.; U Rehman, M.; Alghamdi, B.S.; Zughaibi, T.A.; Ashraf, G.M. Involvement of Phytochemical-Encapsulated Nanoparticles' Interaction with Cellular Signalling in the Amelioration of Benign and Malignant Brain Tumours. *Molecules,* **2022,** *27*(11), 3561.
[http://dx.doi.org/10.3390/molecules27113561] [PMID: 35684498]

[60]   Zhang, J.; Yang, C.; Wei, D.; Li, H.; Leung, E.L.H.; Deng, Q.; Liu, Z.; Fan, X.X.; Zhang, Z. Long-

term efficacy of Chinese medicine Bushen Capsule on cognition and brain activity in patients with amnestic mild cognitive impairment. *Pharmacol. Res.,* **2019**, *146*, 104319.
[http://dx.doi.org/10.1016/j.phrs.2019.104319] [PMID: 31220560]

[61]   Zhang, L.; Yang, S.; Wong, L.R.; Xie, H.; Ho, P.C.L. *In vitro* and *in vivo* comparison of curcumin-encapsulated chitosan-coated poly (lactic- co-glycolic acid) nanoparticles and curcumin/Hydroxypropyl-β-Cyclodextrin inclusion complexes administered intranasally as therapeutic strategies for Alzheimer's diseas. *Mol. Pharm.,* **2020**, *17*(11), 4256-4269.
[http://dx.doi.org/10.1021/acs.molpharmaceut.0c00675] [PMID: 33084343]

[62]   Raju, M.; Kunde, S.S.; Auti, S.T.; Kulkarni, Y.A.; Wairkar, S. Berberine loaded nanostructured lipid carrier for Alzheimer's disease: Design, statistical optimization and enhanced *in vivo* performance. *Life Sci.,* **2021**, *285*, 119990.
[http://dx.doi.org/10.1016/j.lfs.2021.119990] [PMID: 34592234]

[63]   Nie, T.; He, Z.; Zhu, J.; Chen, K.; Howard, G.P.; Pacheco-Torres, J.; Minn, I.; Zhao, P.; Bhujwalla, Z.M.; Mao, H-Q.; Liu, L.; Chen, Y. Non-invasive delivery of levodopa-loaded nanoparticles to the brain *via* lymphatic vasculature to enhance treatment of Parkinson's disease. *Nano Res.,* **2021**, *14*(8), 2749-2761.
[http://dx.doi.org/10.1007/s12274-020-3280-0]

[64]   Deng, J.; Wu, Z.; Zhao, Z.; Wu, C.; Yuan, M.; Su, Z.; Wang, Y.; Wang, Z. Berberine-loaded nanostructured lipid carriers enhance the treatment of ulcerative colitis. *Int. J. Nanomedicine,* **2020**, *15*, 3937-3951.
[http://dx.doi.org/10.2147/IJN.S247406] [PMID: 32581538]

[65]   van Vliet, E.F.; Knol, M.J.; Schiffelers, R.M.; Caiazzo, M.; Fens, M.H.A.M. Levodopa-loaded nanoparticles for the treatment of Parkinson's disease. *J. Control. Release,* **2023**, *360*, 212-224.
[http://dx.doi.org/10.1016/j.jconrel.2023.06.026] [PMID: 37343725]

[66]   Kumbhar, S.; Khairate, R.; Bhatia, M.; Choudhari, P.; Gaikwad, V. Evaluation of curcumin-loaded chitosan nanoparticles for wound healing activity. *ADMET DMPK,* **2023**, *11*(4), 601-613.
[http://dx.doi.org/10.5599/admet.1897] [PMID: 37937244]

[67]   Cardoso, M.A.; Mogharbel, B.F.; Irioda, A.C.; Stricker, P.E.F.; Slompo, R.C.; Perussolo, M.C.; Travelet, C.; Halila, S.; Borsali, R.; de Carvalho, K.A.T. Biodegradable nanoparticles loaded with levodopa and/or curcumin for treatment of Parkinson's disease. *Eur. J. Public Health,* **2021**, *31* Suppl. 2, ckab120.070.
[http://dx.doi.org/10.1093/eurpub/ckab120.070]

[68]   Siafaka, P.I.; Okur, M.E.; Erim, P.D.; Çağlar, E.Ş.; Özgenç, E.; Gündoğdu, E.; Köprülü, R.E.P.; Karantas, I.D.; Üstündağ Okur, N. Protein and Gene Delivery Systems for Neurodegenerative Disorders: Where Do We Stand Today? *Pharmaceutics,* **2022**, *14*(11), 2425.
[http://dx.doi.org/10.3390/pharmaceutics14112425] [PMID: 36365243]

[69]   Gambaryan, P.Y.; Kondrasheva, I.G.; Severin, E.S.; Guseva, A.A.; Kamensky, A.A. Increasing the Effciency of Parkinson's Disease Treatment Using a poly(lactic-co-glycolic acid) (PLGA) Based L-DOPA Delivery System. *Exp. Neurobiol.,* **2014**, *23*(3), 246-252.
[http://dx.doi.org/10.5607/en.2014.23.3.246] [PMID: 25258572]

[70]   Mogharbel, B.F.; Cardoso, M.A.; Irioda, A.C.; Stricker, P.E.F.; Slompo, R.C.; Appel, J.M.; de Oliveira, N.B.; Perussolo, M.C.; Saçaki, C.S.; da Rosa, N.N.; Dziedzic, D.S.M.; Travelet, C.; Halila, S.; Borsali, R.; de Carvalho, K.A.T. Biodegradable Nanoparticles Loaded with Levodopa and Curcumin for Treatment of Parkinson's Disease. *Molecules,* **2022**, *27*(9), 2811.
[http://dx.doi.org/10.3390/molecules27092811] [PMID: 35566173]

[71]   Desai, N. Challenges in development of nanoparticle-based therapeutics. *AAPS J.,* **2012**, *14*(2), 282-295.
[http://dx.doi.org/10.1208/s12248-012-9339-4] [PMID: 22407288]

[72]   Ding, Y.; Kan, J. Optimization and characterization of high pressure homogenization produced

chemically modified starch nanoparticles. *J. Food Sci. Technol.,* **2017**, *54*(13), 4501-4509.
[http://dx.doi.org/10.1007/s13197-017-2934-8] [PMID: 29184257]

[73]   Zhi, K.; Raji, B.; Nookala, A.R.; Khan, M.M.; Nguyen, X.H.; Sakshi, S.; Pourmotabbed, T.; Yallapu, M.M.; Kochat, H.; Tadrous, E.; Pernell, S.; Kumar, S. Plga nanoparticle-based formulations to cross the blood-brain barrier for drug delivery: From r&d to cgmp. *Pharmaceutics,* **2021**, *13*(4), 500.
[http://dx.doi.org/10.3390/pharmaceutics13040500] [PMID: 33917577]

CHAPTER 4

Intranasal Administration for Targeting Neurodegenerative Disorders

Abstract: The Blood-Brain Barrier (BBB) limits the ability of therapeutic molecules to reach the brain following oral or parenteral administration. The nasal delivery system has the potential to be used for drug delivery due to its ease of administration and increased bioavailability. This approach to brain targeting has shown great promise and is useful in treating a range of illnesses linked to dysfunctional brain function. This, along with drug elimination and inactivation during the drug's journey in the systemic circulation and hepatic metabolism, reduces the effectiveness of treatment, necessitates high drug dosages, and frequently results in unfavorable side effects. The anatomical benefits of the nasal route, which allow for the direct delivery of drugs from the nasal cavity to the brain and avoid the blood-brain barrier, are the driving force behind this developing discipline. In addition to playing a significant role in the pathophysiology of neurodegenerative illnesses, oxidative stress can also play a significant role in the damage caused by cerebral ischemia and apoptosis. An interesting new development in medicine is the fascinating intersection of medicinal plants, their bioactive constituents, and nanotechnology, which has shown promise in the treatment of various NDDs. Drug concentration in the brain is increased through nose-to-brain delivery, which circumvents the blood-brain barrier and permits the direct movement of therapeutic molecules.

Keywords: Cerebellomedullary, Nose-to-Brain, Nanoemulsion, Nano phytomedicines.

INTRODUCTION

The field of phytomedicine also known as herbal medicine with therapeutic and healing properties, has been spurred by the growing interest in phytochemicals, that is, chemicals present in fruits, vegetables, nuts, grains, legumes, and other plant foods that have health-promoting effects. Research shows that plant medicine has been used since the conception of disease cure [1]. Even now, Chinese and Indian herbal medicine continues to impact Western medicine. Nonetheless, a return to phytomedicines has been observed in recent years because of their affordability, effectiveness, and reduced potential for adverse

Shivendra Mani Tripathi, Sudhanshu Mishra, Rishabha Malviya & Smriti Ojha

effects. Plant parts such as leaves, stems, bark, roots, and fruits can be utilized to stop, postpone, or reverse the symptoms of a wide range of illnesses [2].

Black tea, *Mucuna pruriens*, *Hibiscus asper* leaves, *Tinospora cordifolia*, sesame seed oil, and *Ginkgo biloba* are among the plants that have been reported to be beneficial in the treatment of various NDs. Numerous beneficial benefits on α-synuclein aggregation, oxidative stress, neuronal degeneration, mitochondrial dysfunction, and locomotor activity have been reported for these and other plants. Researchers studying NDs have been increasingly interested in polyphenols, which are compounds found in plants that have a variety of qualities such as antioxidant, anti-inflammatory, and antiapoptotic effects [3]. Curcumin, ellagic acid, quercetin, and sulforaphane are examples of polyphenols that have similar health benefits to the plants mentioned above. Though their pharmacokinetic qualities still require work, phytomedicines have the potential to be the treatment of choice for various NDs like Parkinson's disease [4]. Certain limitations, including instability, lipophilicity, and molecular size, explain why phytomedicines show promise *in vitro* but less efficacy *in vivo*. Furthermore, the BBB also lessens the *in vivo* efficacy of phytomedicines, which is a concern with many drugs used to treat NDs [5].

By closely controlling the movement of chemicals into the brain, the Blood-Brain Barrier (BBB) divides the central nervous system from the peripheral blood circulation system. It is composed of five different parts: pericytes, astrocyte foot processes, microglia, basement membrane, and endothelial cells of capillaries. Each of these elements contributes special qualities that help maintain the BBB's strict regulation [6]. The absence of fenestrations in endothelial cells restricts the diffusion of proteins and small molecules. Furthermore, the diffusion of compounds soluble in water is inhibited by the interendothelial junctions that exist between the epithelial cells. The BBB is in charge of eliminating waste and keeping the brain's equilibrium while serving as a barrier to keep infections and toxins out. Because of this, only around 2% of small-molecule drugs and even fewer large-molecule treatments can penetrate the blood-brain barrier and reach the brain. This poses a significant obstacle to medication delivery to the brain in the treatment of neurodegenerative illnesses like Parkinson's disease [7]. Therefore, it is critical to examine phytomedicines' poor pharmacokinetic qualities before employing them to treat NDs. This may be accomplished by using carriers like nanoplatforms, along with advanced formulation strategies, which may facilitate the entry of phytochemicals *via* the nose-to-brain route. Certain drugs' physicochemical characteristics prevent them from passing across the blood-brain barrier and from reaching subtherapeutic concentrations in the tissues they are intended to treat. In this regard, the intranasal route presents a viable means of delivering medication to the brain due to its distinct anatomical characteristics.

Specifically, methods based on nanoparticles have proven to have an exceptional ability to get over the difficulties posed by the intranasal route and generate drug accumulation in the brain without going through the system [8].

Nose-to-Brain Drug Delivery Pathway

The nasal-brain administration route has gained interest recently as a viable substitute for intracranial medication delivery. This is because the paths for drug transport from the nasal cavity to the brain are being investigated more thoroughly than in the past as studies on the physiological and neurological systems of the nasal brain are documented [9]. It has recently been suggested that a fresh target for neurological illnesses may be the nasal-brain lymphatic connection. As an appealing substitute for the conventional parenteral and oral methods for the direct delivery of drugs to the brain, intranasal administration has been proposed as a means of achieving high drug levels in the brain. Due to the nasal cavity's special anatomical features, drug delivery is minimally invasive, has a quick stimulation of action, and does not cause a hepatic first-pass effect [10].

In paracellular transport, the cells engaged in this process are called sustentacular cells, and the absorption happens slowly and without the use of energy, while transcellular transport happens across the sustentacular cells and only involves lipophilic medicines [11]. Absorption *via* the systemic route from the respiratory area and the olfactory zone of the nose straight into the brain is one of two important routes for targeting the brain *via* the intranasal route. When a formulation administered *via* the intranasal route comes into touch with the nasal mucosa, it bypasses the blood-brain barrier and enters the brain quickly [12]. Furthermore, the highly hydrophobic medication exhibits effective targeting ability because of its high partition coefficient value.

The nasal cavity receives blood flow from the maxillary, ophthalmic, and facial arteries, thus nasal mucosa is highly vascularized. As a result, drug molecules injected into nasal cavity veins can swiftly travel from the carotid artery to the brain, where they exhibit reverse transfer [13]. Administering drugs by intranasal method comes after the drugs' perivascular transit to the central nervous system. The olfactory region having an olfactory nerve is one of the main pathways for drug absorption. When the drugs are delivered by the intranasal route, they primarily target the brain *via* this channel. The olfactory area of the nose contains supporting cells along with olfactory receptor neurons. The ophthalmic division, maxillary division, and mandibular region of the trigeminal nerve are responsible for transmitting sensory data to the brain from the oral cavity, nasal cavity, eyelids, and cornea. This region collectively forms a trigeminal nerve pathway for drug absorption to the brain *via* the nasal route [14].

Factor Affecting Nose-to-Brain Targeting

Physiological Factor

The primary barriers to nasal drug absorption include efflux transport proteins, blood vessel absorption, enzyme activity, and fast mucociliary clearance. Mucociliary clearance is a protective mechanism that reduces absorption efficiency by limiting drug contact time in the nasal cavity to 20–30 minutes. Cilia transport mucus at a rate of 5–6 mm/min [15]. Nanoparticles, liposomes, and microemulsions are examples of modified drug carriers or bioadhesive materials that can be utilized to lengthen medication adherence and absorption time to combat this. Transport proteins such as P-glycoprotein are involved in efflux processes, which further restrict drug absorption. P-gp inhibitors, for example, can improve medication absorption into the brain by inhibiting these proteins [16]. Drugs are disintegrated down by nasal mucosa enzymes, including proteases and cytochrome subunits, resulting in a "pseudo-first pass effect" that mostly affects peptide- and protein-based drugs. This deterioration can be minimized by enzyme inhibitors. Furthermore, the capillaries in the nasal cavity aid in the absorption of drugs into the circulation, which lessens their direct transport to the brain and increases their negative effects. Vasoconstrictors can help improve brain-targeted delivery by limiting this absorption. Improving these variables can increase the efficacy of nasal drug administration [17].

Chemical and Physical Properties of the Drug

Higher molecular weight normally results in less effective drug absorption. Nasal mucosa absorption of lipid-soluble drugs is over 100% but decreases dramatically as molecular weight rises above 1 kDa. When the molecular weight of macromolecular water-soluble drugs rises, the CSF drug concentration falls. When a polar drug's molecular weight is less than 300 Da, its absorption is mostly unaffected by its physical and chemical characteristics; nevertheless, absorption is greatly influenced when it surpasses 300 Da [18]. A drug must dissolve in nasal secretions, which are 90% water before it can be absorbed *via* the nasal mucosa. Drugs that are highly soluble in water easily dissolve in nasal secretions. Low-solubility drugs can be made more soluble in mucus by using carriers such as cyclodextrins, microemulsions, and nanoparticles. High liposolubility makes drugs easier to absorb *via* the mucosal barrier, but it also increases the risk of entering the bloodstream and decreasing brain targeting [19]. Drug contact with the nasal mucosa is improved by appropriate viscosity, which increases absorption effectiveness. However, as metoclopramide hydrochloride illustrates, higher viscosity might result in less effective absorption. Viscosity is balanced by new carriers such as microspheres, gels, and nanoparticles, which increase retention and absorption efficiency [20].

Formulation Factor

The drug's dissolution, absorption, and penetration are influenced by the pH of nasal preparations and the surface makeup of the nasal cavity; the best absorption occurs when the drug is not dissipative. The pH of nasal secretions is around 5.5–6.5, thus to prevent irritation, preparations should ideally maintain a pH of 4.5–6.5 [21]. Variations in osmotic pressure influence the contraction or relaxation of nasal epithelial cells, which may result in mucosal edema and a decrease in the frequency of ciliated cell beats. Nasal drugs should be isotonic with nasal mucosa, meaning they should be comparable to a 0.9% solution of sodium chloride. Nasal drops, sprays, powders, gels, microspheres, membranes, emulsions, liposomes, nanoparticles, and micelles are among the frequently used formulations for administering drugs through the nose. These formulations' physical characteristics affect how well the drugs dissolve and stay in the nasal cavity. Regarding bioavailability, nasal sprays are superior to nasal drops since the latter are quickly removed by cilia. Sprays improve absorption and bioavailability, lengthen retention times, and primarily deposit drugs in the front part of the nasal cavity [22]. Stronger, longer mucosal contact is provided by powder-based formulations, which boosts drug absorption and brain bioavailability. Nasal medication absorption is also enhanced by other dosage forms such as gels, microspheres, emulsions, liposomes, nanoparticles, and micelles [23].

Experimental Techniques for Intranasal Drug Delivery

Since *in vitro* methods are inadequate for assessing brain-targeted drug delivery, intranasal drug delivery research uses a variety of experimental methodologies, mostly based on pharmacokinetic or pharmacodynamic techniques. The cerebellomedullary cistern puncture approach requires a large number of animals and yields little information on how drug concentrations vary over time [24]. It includes taking Cerebrospinal Fluid (CSF) from mice after administering drugs. Another popular technique, brain tissue homogenization, produces precise time-point data on drug dispersion but requires a large sample size to account for individual variability. The radionuclide labeling technique uses isotope labeling to quantify the amount of drugs in tissues. It provides quick detection but is unable to distinguish between different drug forms [25]. Although it requires expensive and specialized equipment, the brain microdialysis approach yields high-resolution spatial and temporal data on free drug concentrations in the central nervous system. Finally, when direct measurement of drug concentration is difficult, the pharmacodynamic assessment approach evaluates pharmacological effects to indirectly determine drug brain absorption through the nasal mucosa. To meet the various demands and goals of research, each approach has certain benefits and drawbacks when it comes to understanding the dynamics of intranasal administration of drugs [26] (Fig. **1**).

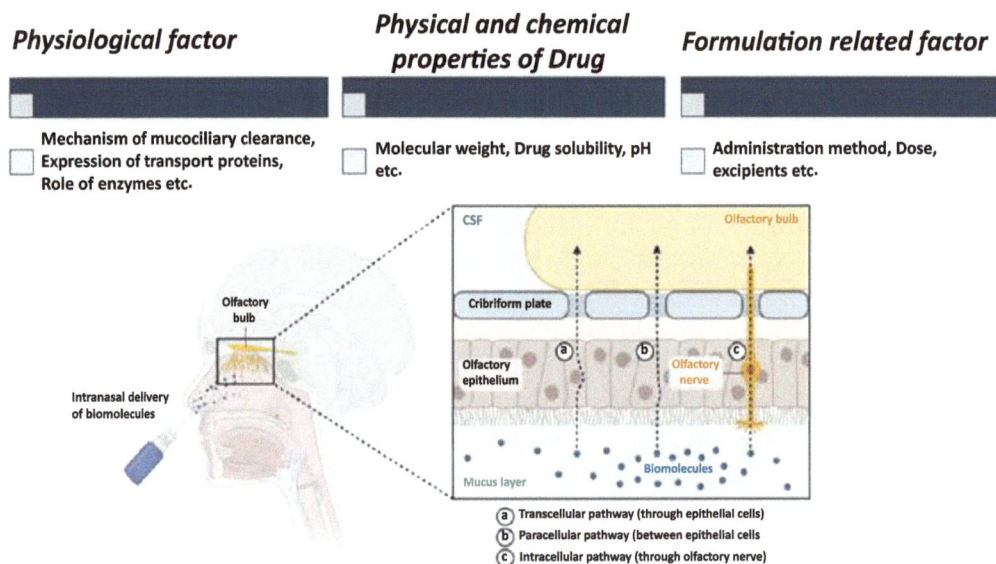

Fig. (1). Factor for intranasal administration of drug.

Formulation Strategies for Nose to Brain Delivery of Phytoconstituents

The most effective method for delivering drugs to the brain is through the non-invasive, patient-friendly nose-to-brain pathway, which avoids the blood-brain barrier. Neurological illnesses in people and animals are targeted *via* the intranasal pathway (olfactory and trigeminal nerves), which circumvents the hepatic first-pass effect, distributes a wide range of hydrophilic and hydrophobic medicines, and permits the entry of many bioactive molecules [27].

A viable substitute for delivering drugs through the nasal route is the creation of advanced drug delivery routes based on nanotechnology, which can even deliver drugs that are cleavage-susceptible and macromolecules [28]. The utilization of nanotechnology in modified delivery has garnered significant interest as a potential solution to address issues related to nasal cavity bioavailability and compliance, which are contingent on the drug's physicochemical characteristics and the physiological parameters of the human nose. Furthermore, target-oriented delivery-specific therapy has several advantages in the treatment of chronic human diseases [29].

This biphasic system is a viable vehicle for the incorporation or encapsulation of several bioactive, nutraceutical, and medicinal substances for both conventional and regulated effects. The two main approaches used to formulate nanoemulsions are high-energy (ultrasonication, microfluidic, high-pressure homogenization) and low-energy (phase inversion) [30].

While the high-energy methods rely on energetic mechanical tactics and disruptive forces to produce smaller oil globules, the low-energy method is based on phase inversion temperature and results in a smaller globular dimension with less energy involved [31].

The lipidic component (5–20%) is used for the o/w kind of nanoemulsion formulation to target the brain. When choosing an oil, solubility is crucial since the drug's molecule needs to be dissolved for the oil to work effectively at the brain's intended place. Lipids such as coconut oil, sesame oil, soybean oil, cottonseed oil, and so on are generally preferable to be consumed alone rather than together [32].

The pharmaceutical business has seen a transformation in recent years due to advancements in nanotechnology and nanomedicine, which have resulted in the creation of innovative drug delivery systems that address the limitations of conventional drug delivery methods [33]. Due to their many benefits and uses, second-generation lipid nanocarriers, or nanostructured lipid carriers, are one such effective and targeted drug delivery system that has drawn a lot of attention worldwide. Although scientific discoveries have transformed our healthcare system, neurological disorders such as brain tumors continue to pose a serious threat because of their short survival rates and difficult treatment delivery to the brain's tissue [34].

Nano Phytomedicines for Nose-to-Brain Drug Delivery

The functional features of nanoparticle-based systems, which are connected to the nanoscale scale and material composition, have led to substantial advancements in the search for ways to deliver therapeutic medicines to the central nervous system. As a result, among other qualities, their surface area, reactivity, strength, sensitivity, and solubility enable them to pass through the BBB [35]. There are three possible ways to deliver nanoparticle-based systems to the central nervous system: non-invasive, invasive, and alternate. The non-invasive techniques rely on endogenous cellular mechanisms that, by colloidal, chemical, or biological features, promote the transport of nanoparticulate-based systems across BBB through the transcellular pathway. The invasive techniques include the use of intraventricular, intrathecal, or interstitial injections to deliver the nanoparticulate-based system directly into the brain tissue. They also include strategies based on disrupting the Blood-Brain Barrier (BBB) such as osmotic, ultrasound, chemical, or magnetic techniques [36].

Linghui and his team produced curcumin-loaded nanoparticles using lactoferrin as a targeting ligand that self-assembled quickly by attaching the hydrophobic portion of lactoferrin to curcumin without the need for hazardous organic chemicals or cross-linking agents. They evaluated the nose-to-brain distribution of developed curcumin-loaded nanoparticles as a targeting ligand as well as a nanocarrier. They also revealed that the developed preparation has significantly demonstrated neuroprotective properties. PC12 cells can internalize these curcumin, favorably loaded NPs, which means that they effectively protect PC12 cells from oxidative stress and apoptosis [37]. The *in vivo* pharmacokinetic findings of their work also showed that these curcumin-loaded NPs can enhance brain accumulation when administered intravenously, as well as extend the elimination half-life and boost the bioavailability of curcumin [38].

Desai and colleagues synthesized curcumin-containing cocrystal micelles intended for intranasal administration as a potential Alzheimer's disease treatment. They revealed that the developed co-crystal has the potential to improve the aqueous solubility of curcumin with enhanced antioxidant potential [39].

Curcumin and chrysin were included in mesoporous silica nanoparticles by Lungare *et al.* for use in nose-to-brain medication delivery. The researchers concluded that these nanocarriers containing chrysin or curcumin may be helpful for antioxidants that are intended to target the brain from the nose [40].

In a recent study by Touitou, E, and co-workers, a nasal nanovesicle delivery system (NVS) for pain management was designed and developed. In an animal model of pain, phospholipid nano vesicular carriers containing ketoprofen, butorphanol, or tramadol produced a speedier onset and enhanced analgesic efficacy [41].

Hao and colleagues created a nasal formulation using an anti-solvent precipitation method that included an ion-activated in situ gel with nanosuspensions loaded with resveratrol. They concluded that resveratrol-containing in situ gel formulation is a potentially effective therapeutic strategy that can lengthen the duration of the drug's residency in the nasal cavity and improve the drug's permeability through the mucosa [42].

Rajput *et al.* developed a nanostructured lipid carrier containing resveratrol by using the melt emulsification–probe sonication technique. A study that carried out the Morris Water Maze test showed that, in contrast to resveratrol suspension taken orally, the drug-loaded lipidic carrier, when supplied through the nose, successfully treats Alzheimer's disease [43]. Emerging applications shown in Fig. (**2**) include integrating gene therapy, stem cell-based approaches, and nanotechnology for enhanced precision and efficacy.

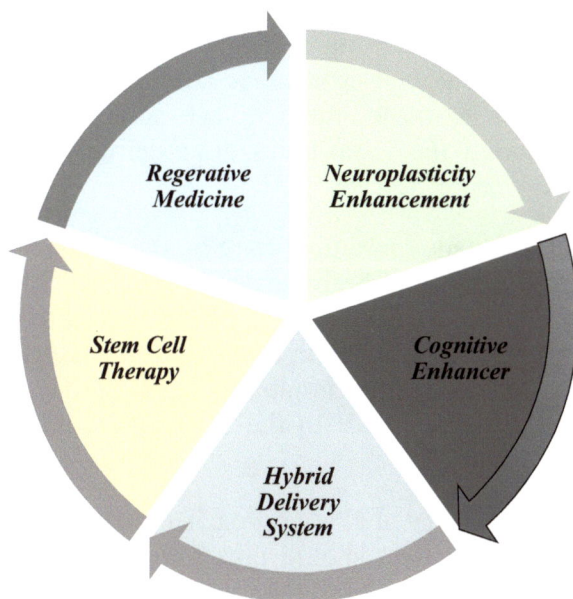

Fig. (2). Potential application for targeting neurological disorder.

Safety Measures for Nose-to-brain Drug Delivery

Although intranasal drug administration is a novel and efficient method of delivering drugs from the nose to the brain, some possible safety issues should be carefully considered. The nasal mucosa and the olfactory nerve system that is situated there may become irritated or injured by frequent and continuous intranasal drug delivery. Intranasal formulations and drugs have the potential of biotoxic effects on the brain in addition to the nasal cavity. The use of permeation enhancers, which promote drug absorption by causing reversible cell growth in the nasal mucosa, has special concerns. Despite improving drug penetration, these enhancers may unintentionally increase the entry of exogenous pathogens, raise the risk of infection, and perhaps cause cell transformation [44]. Furthermore, formulations that contain mucoadhesive ingredients to improve retention in the nasal mucosa may eventually cause toxicity to the mucociliary system. Such toxicity may impair the nasal mucosa's defenses, leading to adverse effects. The long-term impact of nasal-brain drug delivery systems is still poorly understood, even though preclinical and clinical studies have not found any potentially fatal adverse effects after single or short exposures. The need for care is highlighted by the possibility of systemic adverse effects brought on by excessive rapid drug transport to the brain. Such delivery systems may cause drug toxicity, which might have serious repercussions by upsetting the cranial nerve system's immunological homeostasis. Pharmacometric research and formulation studies are essential to addressing these issues. To maintain a balance between therapeutic efficacy and safety, these trials must optimize clinical doses and settings. However, in contrast to more well-established administration pathways like oral or intravenous approaches, these attempts are complicated by the lack of mechanistic clarity around the nasal-brain drug delivery pathway as well as variability in drug delivery efficiency [45].

Resolving these issues is made more important by the therapeutic requirement of efficiently delivering drugs to the brain, especially for the treatment of diseases like encephalopathy. During traditional delivery, the Blood-Brain Barrier (BBB) frequently restricts drug concentrations in brain tissues, which reduces therapeutic results. According to recent studies, intranasal delivery may be able to directly transfer drugs from the nasal cavity to the brain by avoiding the blood-brain barrier [46].

A more thorough risk analysis of this approach is necessary, nevertheless. To reduce these dangers, researchers should take into account elements including immunological reactions, medication toxicity, and long-term nasal mucosal injury. The safety profile of intranasal drug delivery systems might be significantly improved by approaches including the development of safer permeation enhancers, measuring equipment for the health of the nasal mucosa, and dosage optimization strategies [47]. Some clinical trial studies associated with the neurodegenerative disease are listed in the following Table **1**.

Table 1. Clinical trials associated with neurodegenerative diseases.

S. No.	Drugs	Disease	Phase	Clinical Trials.gov ID	References
1.	Insulin	Parkinson's Disease	Phase 2	NCT04251585	[48]
2.	Glutathione	Parkinson's Disease	Phase 1	NCT01398748	[49]
3.	Levodopa	Parkinson's Disease	Phase 2	NCT03541356	[50]
4.	Insulin glulisine	Alzheimer's Disease	Phase 2	NCT02503501	[51]
5.	Oxytocin	Pick's disease.	Phase 2	NCT03260920	[52]
6.	Insulin Aspart	Early Alzheimer's Disease	Phase 1 Phase 2	NCT00581867	[53]
7.	Intranasal auto-M--BFs	Neurocognitive Disorders	Phase 1 Phase 2	NCT02957123	[50]

(Table 1) cont.....

S. No.	Drugs	Disease	Phase	Clinical Trials.gov ID	References
8.	Insulin and Glutathione	Parkinson's Disease	Phase 2	NCT05266417	[54]

Challenges and Limitation

The aforementioned synopses emphasize the benefits of utilizing the promising new wave of nano phytomedicines. Based on the NP mechanism of action, these nanophytomedicines can be divided into three primary categories: targeted drug delivery, device-assisted drug delivery, and drug delivery carriers. The bulk of research conducted thus far has focused on drug delivery carriers, as the name implies. The group that deals with device-assisted medication delivery employs laser irradiation (NIR) or ultrasound to improve the effectiveness of nanophytomedicines. The final category, tailored drug delivery, is under the purview of precision medicine. To guarantee that the phytomedicine crosses the blood-brain barrier, ligands like lactoferrin and rabies virus glycoprotein are used [55].

Numerous synthetic methods, including chitosan hydrogels in situ, chitosan nanoparticles, and mucoadhesive nanostructured lipid carriers, have been employed to get beyond the nasal mucosa barrier and boost the effectiveness of medication therapy. Nevertheless, a number of these methods could result in membrane component leaching and nasal toxicity, which would irritate the nasal mucosa locally [56]. Among the most promising methods of delivering therapeutic chemicals to the central nervous system is the intranasal administration of nanoparticulate-based systems, notwithstanding all the obstacles encountered. Taken combined, the benefits of the delivery method and the characteristics of the nanoparticles can make it easier to deliver drugs to the central nervous system [57]. Thus, one of the primary parameters that need to be regulated in the creation of intranasal formulations loaded with phytochemicals is the size of nanoparticulate-based systems. Particle size of particular phytochemicals can also affect drug loading, release, and stability, which in turn affects toxicity, *in vivo* distribution, and CNS targeting ability [58]. The pharmacokinetics of nanocarriers, such as the circulation time, absorption, and biodistribution, can also be influenced by the particle size distribution. More surface area and smaller particle sizes can therefore result in better drug solubility, stronger mucosal interactions, or better penetration than a drug solution, which may even be favored by the kind of nanoparticulate system composition. Improved medication performance following administration can also be attributed to the surface charge of nanocarriers [59]. To favor the formulation's longer-term retention in the nasal mucosa, positive zeta potentials may facilitate greater interaction with the negatively charged mucin residues. Despite this, some studies showed that administering artificial nanovesicles *via* the nose offered notable benefits in terms of safety and transport effectiveness for treating brain diseases. BBB as a whole is thought to be the primary obstacle preventing drugs from penetrating the central nervous system. Because intranasal administration circumvents the BBB more successfully than systemic administration, it presents an alluring substitute for central nervous system administration [60]. Intranasal administration, despite its advantages, has limitations (Fig. **3**) such as rapid mucociliary clearance, enzymatic degradation in the nasal mucosa, and potential irritation with frequent use. Variability in individual anatomy can affect drug absorption, while systemic absorption may reduce brain targeting and increase side effects.

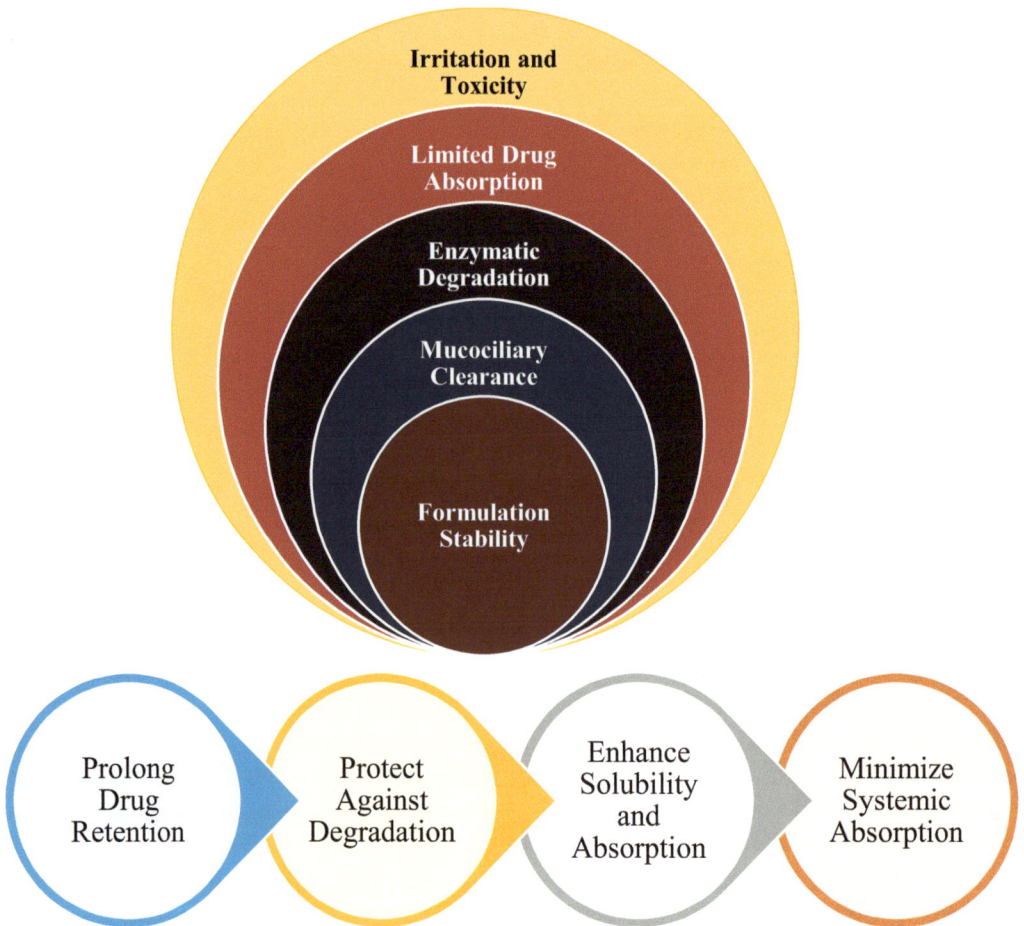

Fig. (3). Limitation and potential solution of intranasal administration.

CONCLUSION

When compared to traditional drug delivery methods, the nose-to-brain system offers better efficacy, effective permeability, enhanced retention, absorption, and bioavailability, as well as a significantly lower risk of side effects. This makes it an advantageous treatment option for Alzheimer's disease. It is anticipated that new nasal products, such as drugs for chronic illnesses, novel nasal vaccines, and drugs that directly target the brain and guarantee a therapeutic effect on the central nervous system with fewer side effects, will continue to hit the market given the broad interest in nasal drug delivery and its potential benefits. As a result, this developing industry will see a great deal of research and development of new technologies in the upcoming years.

REFERENCES

[1]   Wachtel-Galor, S.; Benzie, I.F.F. *Herbal medicine: An introduction to its history, usage, regulation, current trends, and research needs. Herb. Med. Biomol. Clin. Asp,* 2nd ed; CRC Press/Taylor & Francis, **2011**, pp. 1-10.

[2]   Li, C.; Jia, W.; Yang, J.; Cheng, C.; Olaleye, O.E. Multi-compound and drug-combination pharmacokinetic research on Chinese herbal medicines. *Acta Pharmacol. Sin.,* **2022**, *43*(12), 3080-3095.
[http://dx.doi.org/10.1038/s41401-022-00983-7] [PMID: 36114271]

[3]   Rocha, E.M.; De Miranda, B.; Sanders, L.H. Alpha-synuclein: Pathology, mitochondrial dysfunction and neuroinflammation in Parkinson's disease. *Neurobiol. Dis.,* **2018**, *109*(Pt B), 249-257.
[http://dx.doi.org/10.1016/j.nbd.2017.04.004] [PMID: 28400134]

[4]   Pandey, K.B.; Rizvi, S.I. Plant polyphenols as dietary antioxidants in human health and disease. *Oxid. Med. Cell. Longev.,* **2009**, *2*(5), 270-278.
[http://dx.doi.org/10.4161/oxim.2.5.9498] [PMID: 20716914]

[5]   Nsairat, H.; Lafi, Z.; Al-Sulaibi, M.; Gharaibeh, L.; Alshaer, W. Impact of nanotechnology on the oral delivery of phyto-bioactive compounds. *Food Chem.,* **2023**, *424*, 136438.
[http://dx.doi.org/10.1016/j.foodchem.2023.136438] [PMID: 37244187]

[6]   Pandit, R.; Chen, L.; Götz, J. The blood-brain barrier: Physiology and strategies for drug delivery. *Adv. Drug Deliv. Rev.,* **2020**, *165-166*, 1-14.
[http://dx.doi.org/10.1016/j.addr.2019.11.009] [PMID: 31790711]

[7]   Niazi, S.K. Non-Invasive Drug Delivery across the Blood–Brain Barrier: A Prospective Analysis. *Pharmaceutics,* **2023**, *15*(11), 2599.
[http://dx.doi.org/10.3390/pharmaceutics15112599] [PMID: 38004577]

[8]   Formica, M.L.; Real, D.A.; Picchio, M.L.; Catlin, E.; Donnelly, R.F.; Paredes, A.J. On a highway to the brain: A review on nose-to-brain drug delivery using nanoparticles. *Appl. Mater. Today,* **2022**, *29*, 101631.
[http://dx.doi.org/10.1016/j.apmt.2022.101631]

[9]   Bonferoni, M.C.; Rossi, S.; Sandri, G.; Ferrari, F.; Gavini, E.; Rassu, G.; Giunchedi, P. Nanoemulsions for "nose-to-brain" drug delivery. *Pharmaceutics,* **2019**, *11*(2), 84.
[http://dx.doi.org/10.3390/pharmaceutics11020084] [PMID: 30781585]

[10]   Jeong, S.H.; Jang, J.H.; Lee, Y.B. Drug delivery to the brain *via* the nasal route of administration: exploration of key targets and major consideration factors. *J. Pharm. Investig.,* **2023**, *53*(1), 119-152.
[http://dx.doi.org/10.1007/s40005-022-00589-5] [PMID: 35910081]

[11]   Yu, A.S.L. Paracellular transport as a strategy for energy conservation by multicellular organisms? *Tissue Barriers,* **2017**, *5*(2), e1301852.
[http://dx.doi.org/10.1080/21688370.2017.1301852] [PMID: 28452575]

[12]   Ruigrok, M.J.R.; de Lange, E.C.M. Emerging Insights for Translational Pharmacokinetic and Pharmacokinetic-Pharmacodynamic Studies: Towards Prediction of Nose-to-Brain Transport in Humans. *AAPS J.,* **2015**, *17*(3), 493-505.
[http://dx.doi.org/10.1208/s12248-015-9724-x] [PMID: 25693488]

[13]   Chung, S.; Peters, J.M.; Detyniecki, K.; Tatum, W.; Rabinowicz, A.L.; Carrazana, E. The nose has it: Opportunities and challenges for intranasal drug administration for neurologic conditions including seizure clusters. *Epilepsy Behav. Rep.,* **2023**, *21*, 100581.
[http://dx.doi.org/10.1016/j.ebr.2022.100581] [PMID: 36636458]

[14]   Keller, L.A.; Merkel, O.; Popp, A. Intranasal drug delivery: opportunities and toxicologic challenges during drug development. *Drug Deliv. Transl. Res.,* **2022**, *12*(4), 735-757.
[http://dx.doi.org/10.1007/s13346-020-00891-5] [PMID: 33491126]

[15]   Marttin, E.; Schipper, N.G.M.; Verhoef, J.C.; Merkus, F.W.H.M. Nasal mucociliary clearance as a factor in nasal drug delivery. *Adv. Drug Deliv. Rev.,* **1998**, *29*(1-2), 13-38.
[http://dx.doi.org/10.1016/S0169-409X(97)00059-8] [PMID: 10837578]

[16]   Taha, M.S.; Nocera, A.; Workman, A.; Amiji, M.M.; Bleier, B.S. P-glycoprotein inhibition with verapamil overcomes mometasone resistance in Chronic Sinusitis with Nasal Polyps. *Rhinology,* **2021**, *0*(0), 0.
[http://dx.doi.org/10.4193/Rhin20.551] [PMID: 33459729]

[17]   Sarkar, M.A. Drug metabolism in the nasal mucosa. *Pharm. Res.,* **1992**, *9*(1), 1-9.
[http://dx.doi.org/10.1023/A:1018911206646] [PMID: 1589391]

[18]   Huang, Q.; Chen, X.; Yu, S.; Gong, G.; Shu, H. Research progress in brain-targeted nasal drug delivery. *Front. Aging Neurosci.,* **2024**, *15*, 1341295.
[http://dx.doi.org/10.3389/fnagi.2023.1341295] [PMID: 38298925]

[19]   Tai, J.; Han, M.; Lee, D.; Park, I.H.; Lee, S.H.; Kim, T.H. Different Methods and Formulations of Drugs and Vaccines for Nasal Administration. *Pharmaceutics,* **2022**, *14*(5), 1073.
[http://dx.doi.org/10.3390/pharmaceutics14051073] [PMID: 35631663]

[20]   Zaki, N.M.; Awad, G.A.S.; Mortada, N.D.; Abd ElHady, S.S. Rapid-onset intranasal delivery of metoclopramide hydrochloride. *Int. J. Pharm.,* **2006**, *327*(1-2), 89-96.
[http://dx.doi.org/10.1016/j.ijpharm.2006.07.040] [PMID: 16942844]

[21]   Pires, P.C.; Rodrigues, M.; Alves, G.; Santos, A.O. Strategies to Improve Drug Strength in Nasal Preparations for Brain Delivery of Low Aqueous Solubility Drugs. *Pharmaceutics,* **2022**, *14*(3), 588.
[http://dx.doi.org/10.3390/pharmaceutics14030588] [PMID: 35335964]

[22]   Zhan, J.; Ting, X.L.; Zhu, J. The Research Progress of Targeted Drug Delivery Systems. **2017**.
[http://dx.doi.org/10.1088/1757-899X/207/1/012017]

[23]   Kaur, G.; Goyal, J.; Behera, P.K.; Devi, S.; Singh, S.K.; Garg, V.; Mittal, N. Unraveling the role of chitosan for nasal drug delivery systems: A review. *Carbohydrate Polymer Technologies and Applications,* **2023**, *5*, 100316.
[http://dx.doi.org/10.1016/j.carpta.2023.100316]

[24]   Samaranch, L.; Bringas, J.; Pivirotto, P.; Sebastian, W.S.; Forsayeth, J.; Bankiewicz, K. Cerebellomedullary cistern delivery for AAV-based gene therapy: A technical note for nonhuman primates. *Hum. Gene Ther. Methods,* **2016**, *27*(1), 13-16.
[http://dx.doi.org/10.1089/hgtb.2015.129] [PMID: 26757202]

[25]   Haasbroek-Pheiffer, A.; Van Niekerk, S.; Van der Kooy, F.; Cloete, T.; Steenekamp, J.; Hamman, J. *In vitro* and *ex vivo* experimental models for evaluation of intranasal systemic drug delivery as well as direct nose☐to☐brain drug delivery. *Biopharm. Drug Dispos.,* **2023**, *44*(1), 94-112.

[http://dx.doi.org/10.1002/bdd.2348] [PMID: 36736328]

[26] Shannon, R.J.; Carpenter, K.L.H.; Guilfoyle, M.R.; Helmy, A.; Hutchinson, P.J. Cerebral microdialysis in clinical studies of drugs: pharmacokinetic applications. *J. Pharmacokinet. Pharmacodyn.,* **2013**, *40*(3), 343-358.
[http://dx.doi.org/10.1007/s10928-013-9306-4] [PMID: 23468415]

[27] Trevino, J.T.; Quispe, R.C.; Khan, F.; Novak, V. Non-Invasive Strategies for Nose-to-Brain Drug Delivery. *J. Clin. Trials,* **2020**, *10*(7), 10.
[PMID: 33505777]

[28] Yue, X.; Guo, H.; Wang, G.; Li, J.; Zhai, Z.; Wang, Z.; Wang, W.; Zhao, Z.; Xia, X.; Chen, C.; Cui, Y.; Wu, C.; Huang, Z.; Zhang, X. A tailored phytosomes based nose-to-brain drug delivery strategy: Silver bullet for Alzheimer's disease. *Bioact. Mater.,* **2025**, *44*, 97-115.
[http://dx.doi.org/10.1016/j.bioactmat.2024.09.039]

[29] Afzal, O.; Altamimi, A.S.A.; Nadeem, M.S.; Alzarea, S.I.; Almalki, W.H.; Tariq, A.; Mubeen, B.; Murtaza, B.N.; Iftikhar, S.; Riaz, N.; Kazmi, I. Nanoparticles in Drug Delivery: From History to Therapeutic Applications. *Nanomaterials (Basel),* **2022**, *12*(24), 4494.
[http://dx.doi.org/10.3390/nano12244494] [PMID: 36558344]

[30] Preeti, S.S.; Sambhakar, S.; Malik, R.; Bhatia, S.; Al Harrasi, A.; Rani, C.; Saharan, R.; Kumar, S.; Geeta, ; Sehrawat, R. Nanoemulsion: An Emerging Novel Technology for Improving the Bioavailability of Drugs. *Scientifica (Cairo),* **2023**, *2023*, 1-25.
[http://dx.doi.org/10.1155/2023/6640103] [PMID: 37928749]

[31] Espinoza, L.C.; Silva-Abreu, M.; Clares, B.; Rodríguez-Lagunas, M.J.; Halbaut, L.; Cañas, M.A.; Calpena, A.C. Formulation strategies to improve nose-to-brain delivery of donepezil. *Pharmaceutics,* **2019**, *11*(2), 64.
[http://dx.doi.org/10.3390/pharmaceutics11020064] [PMID: 30717264]

[32] Azeem, A.; Rizwan, M.; Ahmad, F.J.; Iqbal, Z.; Khar, R.K.; Aqil, M.; Talegaonkar, S. Nanoemulsion components screening and selection: a technical note. *AAPS PharmSciTech,* **2009**, *10*(1), 69-76.
[http://dx.doi.org/10.1208/s12249-008-9178-x] [PMID: 19148761]

[33] Farjadian, F.; Ghasemi, A.; Gohari, O.; Roointan, A.; Karimi, M.; Hamblin, M.R. Nanopharmaceuticals and nanomedicines currently on the market: challenges and opportunities. *Nanomedicine (Lond.),* **2019**, *14*(1), 93-126.
[http://dx.doi.org/10.2217/nnm-2018-0120] [PMID: 30451076]

[34] Misra, S.K.; Pathak, K. Nose-to-Brain Targeting *via* Nanoemulsion: Significance and Evidence. *Colloids and Interfaces,* **2023**, *7*(1), 23.
[http://dx.doi.org/10.3390/colloids7010023]

[35] Garg, J.; Pathania, K.; Sah, S.P.; Pawar, S.V. Nanostructured lipid carriers: a promising drug carrier for targeting brain tumours. *Future Journal of Pharmaceutical Sciences,* **2022**, *8*(1), 25.
[http://dx.doi.org/10.1186/s43094-022-00414-8]

[36] Jitta, S.R.; Kumar, L. *Nanostructured lipid carriers as an alternative carrier with high drug loading for targeting to brain tumors*; Nanocarriers for Drug-Targeting Brain Tumors, **2022**, pp. 269-297.
[http://dx.doi.org/10.1016/B978-0-323-90773-6.00015-4]

[37] Bicker, J.; Fortuna, A.; Alves, G.; Falcão, A. Nose-to-brain Delivery of Natural Compounds for the Treatment of Central Nervous System Disorders. *Curr. Pharm. Des.,* **2020**, *26*(5), 594-619.
[http://dx.doi.org/10.2174/1381612826666200115101544] [PMID: 31939728]

[38] Li, L.; Tan, L.; Zhang, Q.; Cheng, Y.; Liu, Y.; Li, R.; Hou, S. Nose-to-brain delivery of self-assembled curcumin-lactoferrin nanoparticles: Characterization, neuroprotective effect and *in vivo* pharmacokinetic study. *Front. Bioeng. Biotechnol.,* **2023**, *11*, 1168408.
[http://dx.doi.org/10.3389/fbioe.2023.1168408] [PMID: 37051277]

[39] Desai, P.P.; Patravale, V.B. Curcumin Cocrystal Micelles—Multifunctional Nanocomposites for

Management of Neurodegenerative Ailments. *J. Pharm. Sci.,* **2018**, *107*(4), 1143-1156.
[http://dx.doi.org/10.1016/j.xphs.2017.11.014] [PMID: 29183742]

[40]　Lungare, S.; Hallam, K.; Badhan, R.K.S. Phytochemical-loaded mesoporous silica nanoparticles for nose-to-brain olfactory drug delivery. *Int. J. Pharm.,* **2016**, *513*(1-2), 280-293.
[http://dx.doi.org/10.1016/j.ijpharm.2016.09.042] [PMID: 27633279]

[41]　Touitou, E.; Natsheh, H.; Boukeileh, S.; Awad, R. Short onset and enhanced analgesia following nasal administration of non-controlled drugs in nanovesicular systems. *Pharmaceutics,* **2021**, *13*(7), 978.
[http://dx.doi.org/10.3390/pharmaceutics13070978] [PMID: 34203555]

[42]　Hao, J.; Zhao, J.; Zhang, S.; Tong, T.; Zhuang, Q.; Jin, K.; Chen, W.; Tang, H. Fabrication of an ionic-sensitive in situ gel loaded with resveratrol nanosuspensions intended for direct nose-to-brain delivery. *Colloids Surf. B Biointerfaces,* **2016**, *147*, 376-386.
[http://dx.doi.org/10.1016/j.colsurfb.2016.08.011] [PMID: 27566226]

[43]　Rajput, A.; Bariya, A.; Allam, A.; Othman, S.; Butani, S.B. In situ nanostructured hydrogel of resveratrol for brain targeting: *in vitro-in vivo* characterization. *Drug Deliv. Transl. Res.,* **2018**, *8*(5), 1460-1470.
[http://dx.doi.org/10.1007/s13346-018-0540-6] [PMID: 29785574]

[44]　Lofts, A.; Abu-Hijleh, F.; Rigg, N.; Mishra, R.K.; Hoare, T. Using the Intranasal Route to Administer Drugs to Treat Neurological and Psychiatric Illnesses: Rationale, Successes, and Future Needs. *CNS Drugs,* **2022**, *36*(7), 739-770.
[http://dx.doi.org/10.1007/s40263-022-00930-4] [PMID: 35759210]

[45]　Wu, H.; Zhou, Y.; Wang, Y.; Tong, L.; Wang, F.; Song, S.; Xu, L.; Liu, B.; Yan, H.; Sun, Z. Current State and Future Directions of Intranasal Delivery Route for Central Nervous System Disorders: A Scientometric and Visualization Analysis. *Front. Pharmacol.,* **2021**, *12*, 717192.
[http://dx.doi.org/10.3389/fphar.2021.717192] [PMID: 34322030]

[46]　Dighe, S.; Jog, S.; Momin, M.; Sawarkar, S.; Omri, A. Intranasal Drug Delivery by Nanotechnology: Advances in and Challenges for Alzheimer's Disease Management. *Pharmaceutics,* **2023**, *16*(1), 58.
[http://dx.doi.org/10.3390/pharmaceutics16010058] [PMID: 38258068]

[47]　Rai, G.; Gauba, P.; Dang, S. Recent advances in nanotechnology for Intra-nasal drug delivery and clinical applications. *J. Drug Deliv. Sci. Technol.,* **2023**, *86*, 104726.
[http://dx.doi.org/10.1016/j.jddst.2023.104726]

[48]　Study Details | Intranasal Insulin in Parkinson's Disease | ClinicalTrials.gov. ClinicalTrials.gov n.d. https://clinicaltrials.gov/study/NCT02064166

[49]　Intranasal Glutathione in Parkinson's Disease. ClinicalTrials.gov n.d. https://clinicaltrials.gov/study/NCT01398748

[50]　Impel NeuroPharma Inc. Therapeutic potential for intranasal levodopa in Parkinson's disease – OFF reversal. ClinicalTrials.gov n.d. https://clinicaltrials.gov/study/NCT04386041

[51]　Rosenbloom, M.; Barclay, T.R.; Kashyap, B.; Hage, L.; O'Keefe, L.R.; Svitak, A.; Pyle, M.; Frey, W.; Hanson, L.R. A Phase II, Single-Center, Randomized, Double-Blind, Placebo-Controlled Study of the Safety and Therapeutic Efficacy of Intranasal Glulisine in Amnestic Mild Cognitive Impairment and Probable Mild Alzheimer's Disease. *Drugs Aging,* **2021**, *38*(5), 407-415.
[http://dx.doi.org/10.1007/s40266-021-00845-7] [PMID: 33719017]

[52]　Study Details | Intranasal Oxytocin for Frontotemporal Dementia. ClinicalTrials.gov n.d. https://clinicaltrials.gov/study/NCT04182723

[53]　Memory and Insulin in Early Alzheimer's Disease. Cochrane Library n.d. https://www.cochranelibrary.com/central/doi/10.1002/central/CN-01919409/full

[54]　*Intranasal Insulin and Glutathione as an Add-On Therapy in Parkinson's Disease*; NOSE-PD, **2022**.

[55]　Burns, J.; Buck, A.C.; D' Souza, S.; Dube, A.; Bardien, S. Nanophytomedicines as Therapeutic Agents

for Parkinson's Disease. *ACS Omega,* **2023**, *8*(45), 42045-42061.
[http://dx.doi.org/10.1021/acsomega.3c04862] [PMID: 38024675]

[56]   Syed Azhar, S.N.A.; Ashari, S.E.; Zainuddin, N.; Hassan, M. Nanostructured Lipid Carriers-Hydrogels System for Drug Delivery: Nanohybrid Technology Perspective. *Molecules,* **2022**, *27*(1), 289.
[http://dx.doi.org/10.3390/molecules27010289] [PMID: 35011520]

[57]   Islam, S.U.; Shehzad, A.; Ahmed, M.B.; Lee, Y.S. Intranasal delivery of nanoformulations: A potential way of treatment for neurological disorders. *Molecules,* **2020**, *25*(8), 1929.
[http://dx.doi.org/10.3390/molecules25081929] [PMID: 32326318]

[58]   Nguyen, T.T.L.; Maeng, H.J. Pharmacokinetics and Pharmacodynamics of Intranasal Solid Lipid Nanoparticles and Nanostructured Lipid Carriers for Nose-to-Brain Delivery. *Pharmaceutics,* **2022**, *14*(3), 572.
[http://dx.doi.org/10.3390/pharmaceutics14030572] [PMID: 35335948]

[59]   Hoshyar, N.; Gray, S.; Han, H.; Bao, G. The effect of nanoparticle size on *in vivo* pharmacokinetics and cellular interaction. *Nanomedicine (Lond.),* **2016**, *11*(6), 673-692.
[http://dx.doi.org/10.2217/nnm.16.5] [PMID: 27003448]

[60]   Ghadiri, M.; Young, P.M.; Traini, D. Strategies to enhance drug absorption *via* nasal and pulmonary routes. *Pharmaceutics,* **2019**, *11*(3), 113.
[http://dx.doi.org/10.3390/pharmaceutics11030113] [PMID: 30861990]

CHAPTER 5

Blood Brain Barrier and Nanotechnology for Neurodegenerative Disorders

Abstract: The Blood-Brain Barrier (BBB) is a highly selective semipermeable membrane that separates the circulating blood from the brain's extracellular fluid, ensuring CNS homeostasis and protecting the brain from harmful substances while allowing essential nutrients to pass. The BBB, composed of specialized endothelial cells connected by tight junctions, poses a significant challenge for drug delivery to the brain due to its restrictive nature. With the help of nanotechnology and the special qualities of nanomaterials such as their tiny size, biocompatibility, and capacity to increase blood circulation promising solutions to these problems remain. The structure and function of the Blood-Brain Barrier (BBB), the potential of nanotechnology for delivering drugs to the brain, and the application of different nanomaterials, such as polymeric, liposomal, dendrimer, and inorganic nanoparticles, in the treatment of neurological disorders are all covered in this study. Also, the neurotoxicity of nanomaterials and the therapeutic and neuroprotective potential of phytoconstituents are explained.

Keywords: Blood-Brain Barrier, CNS homeostasis, Endothelial cells, Nanomaterials.

INTRODUCTION

The Blood-Brain Barrier (BBB) is a crucial interface between the bloodstream and the brain that plays a vital role in maintaining the homeostasis of the Central Nervous System (CNS). It is a highly selective semipermeable membrane that separates the circulating blood from the brain's extracellular fluid, regulating the passage of molecules, ions, and cells into and out of the brain [1]. The BBB is essential for protecting the brain from harmful substances while allowing the passage of essential nutrients and molecules necessary for brain function. Being the tightest endothelium in the body, the BBB also represents the main impediment to drug delivery to the brain [2]. The vascular tree is comprised of arteries and arterioles, which deliver blood to the tissues, the capillary bed, which is essential for gas and nutrient exchange within tissues, and venules and veins, which drain blood from tissues [3]. Each segment has different properties depending on where they are in the vascular tree as well as which organ they vas-

Shivendra Mani Tripathi, Sudhanshu Mishra, Rishabha Malviya & Smriti Ojha

cularize. There are three main structural classes of capillaries [4]. Continuous no-fenestrated capillaries of the skin and lung are joined together by cellular junctions, have a complete Basement Membrane (BM), and lack fenestra (pores) in their plasma membrane. Continuous fenestrated vessels of the intestinal villi and endocrine glands have a similar continuous structure but contain diaphragmed fenestra throughout their membrane [5].

The advancement of nanotechnology *via* integrated, interdisciplinary efforts will lead to novel perspectives of how neural circuits work as well as methods for the diagnosis and treatment of brain disorders [6]. The shortcomings of existing methods for delivering drugs across the blood-brain barrier to the central nervous system make this particularly important [7]. The unique characteristics of nanomaterials, including their smaller size, biocompatibility, ability to extend blood circulation, and lack of toxicity, have been utilized to develop a novel delivery system that facilitates the effortless administration of therapeutic compounds to the brain [8]. Targeting particular and non-specific brain locations is the foundation of nanotechnology-mediated drug delivery systems [9].

Artificial Intelligence (AI) is advancing nanocarrier formulation for brain targeting *via* the nasal route by enabling predictive modeling and optimization of nanoparticle properties. AI tools, like machine learning algorithms, predict drug encapsulation efficiency, particle size, and biocompatibility, tailoring formulations for better drug release and targeting [10].

These technologies integrate computational models for personalized treatment, leveraging insights from patient-specific pharmacokinetics and imaging data. AI enhances nanocarrier design efficiency and supports rapid prototyping of smart, stimuli-responsive carriers for neurological applications while overcoming biological barriers like the blood-brain barrier [11].

Structure and Function of the BBB

The BBB is primarily formed by specialized endothelial cells that line the capillaries in the brain. These endothelial cells are connected by tight junctions, which are protein complexes that seal the gaps between cells, preventing the passage of large molecules and pathogens [12]. In addition to tight junctions, the BBB is also supported by pericytes, astrocytes, and the basement membrane, all of which play important roles in maintaining the integrity of the barrier [13]. The Blood-Brain Barrier (BBB) is a term used to describe the unique properties of the microvasculature of the Central Nervous System (CNS) [14]. CNS vessels are continuous non-fenestrated vessels but also contain a series of additional properties that allow them to tightly regulate the movement of molecules, ions, and cells between the blood and the CNS [15]. This heavily restricting barrier capacity allows BBB ECs to tightly regulate CNS homeostasis, which is critical to allow for proper neuronal function, as well as protect the CNS from toxins, pathogens, inflammation, injury, and disease [16]. The restrictive nature of the BBB provides an obstacle to drug delivery to the CNS, and, thus, major efforts have been made to generate methods to modulate or bypass the BBB for the delivery of therapeutics [17]. Loss of some, or most, of these barrier properties during neurological diseases including stroke, Multiple Sclerosis (MS), brain traumas, and neurodegenerative disorders, is a major component of the pathology and progression of these diseases [18].

These blood vessels are made up of two main types of cells namely, endothelial cells and mural cells. These endothelial cells are mesodermally derived modified simple squamous epithelial cells that form the walls of blood vessels [19]. Microvascular ECs are extremely thin cells that are 39% less thick than muscle ECs, with a distance of less than a quarter of a micron separating the luminal from the parenchymal surface. For morphology, the endothelial cells in the BBB are fastened by both tight junctions and adherens junctions, resulting in distinct luminal and abluminal membrane compartments [20]. The functions of these ECs are, first, they display a net negative surface charge, refusing to accept negatively charged compounds, as well as quite low degrees of leukocyte adhesion molecules, hampering the entry of the number of immune cells [21]. Second, they show designated transporters for regulating the inflow and outflow of specific substrates. Third, they show a restriction on the number of transcellular vesicles through the vessel wall due to the high transendothelial electrical resistance. Because of the existence of the local environment, endothelial cells can together form and maintain the BBB. Molecules cross the BBB by a paracellular pathway (between adjacent cells) or a transcellular pathway (through the cells) [22]. For the paracellular pathway, ions, and solutes utilize concentration gradients to pass the BBB by passive diffusion. The transcellular pathway includes different mechanisms such as passive diffusion, receptor-mediated transport, and transcytosis [23]. Overall, passive diffusion is a non-saturable mechanism dependent on the physicochemical properties of the molecule. The physicochemical factors that influence BBB permeability include molecular weight, charge, lipid solubility, surface activity, and relative size of the molecule [24]. The BBB performs 3 major functions: it curbs free transport between the blood and the brain, essential nutrients are supplied to the brain through it, and it aids in the flow of any harmful or toxic waste or foreign substances [25]. The brain endothelial cells that form the BBB express transport proteins. These are mostly separated within either the luminal or abluminal surfaces, which mainly give expressions for the transport of peptides, proteins, and neurotransmitter-metabolizing enzymes [26].

Mural Cells Mural cells include vascular smooth muscle cells that surround the large vessels and pericytes, which incompletely cover the endothelial walls of the microvasculature [27]. Pericytes (PCs) are cells that sit on the abluminal surface of the microvascular endothelial tube and are embedded in the vascular BM. A difficulty in studying PCs is the lack of a specific marker that is expressed uniquely by PCs, and, thus, these cells are often confused with other cells that sit in the perivascular space [28]. One of the major questions in pericyte biology is whether different subsets of PCs may have different functions. Owing to the lack of defining markers, it remains unclear whether all of the different functions attributed to PCs are performed by all of the same cells, by different subsets of PCs, or even by non-pericyte cells [29].

Immune cells in blood vessels interact with different immune cell populations both within the blood as well as within the CNS. The two main cell populations in the CNS are perivascular macrophages and microglial cells. Perivascular macrophages are monocyte lineage cells that sit on the abluminal side of the vascular tube [30].

Astrocytes, also known as astroglia, are the most numerous glial cells, expressing polarized and complex morphology, which are heterogeneous throughout the brain. In tradition, they are divided into two categories, one is protoplasmic which is located in the well-vascularized gray matter, and the other one is fibrous, which is located in the less vascular white matter [31, 32]. Their end feet link them with the basement membrane, *via* the binding of a set of proteins (aquaporin IV and the dystroglycan-dystrophin complex) with the proteoglycan agrin. In the CNS, they play a major role in dynamic signaling such as clearing waste, tuning brain blood flow, regulating vascular function, ion hemostasis, and balancing neuroimmune responses [33].

Etiology of Neurodegenerative Disorders

The term "neurodegeneration" refers to the progressive loss of neurons that causes serious neurological disorders such as motor neuron disease, Parkinsonism, and dementia. The main pathogenic characteristics include oxidative stress, aberrant protein aggregation formation, mitochondrial dysfunction, and dysfunctional proteasomes [34]. More than 350,000 Australians suffer from dementia, and 80,000 from Parkinson's disease; these figures are projected to increase as the country's population ages. By 2020, there will likely be 930,000 instances of Parkinson's disease in the US, compared to the 5.4 million cases of Alzheimer's in 2016 [35]. Certain genetic variants that raise the risk of disease have been found, even though genetic factors are poorly understood. Understanding neurodegeneration requires examining postmortem human brains and animal models, with the use of sophisticated bioinformatics and molecular imaging [36]. It is difficult to correlate pathological and clinical results since neurodegeneration is a complicated condition. For advancements in etiology and therapy, a research approach that incorporates future molecular targets, efficient drug transport to the brain, and well-planned clinical trials is essential [37].

Blood Brain Barrier and Nanotechnology

The Blood-Brain Barrier (BBB) is an essential defense system that controls the environment in the brain to maintain neuronal function and keep harmful substances outside. The Blood-Brain Barrier (BBB), which is made up of endothelial cells linked by Tight Junctions (TJs), blocks molecule transit while permitting lipid-soluble substances to flow through readily [38]. The BBB effectively blocks the passage of 95% of drugs, making drug delivery to the Central Nervous System (CNS) very difficult. Traditional drug administration techniques frequently fall short of efficiently penetrating the BBB [39]. Lipid-soluble materials can move readily, however, many drugs have trouble getting to the therapeutic levels because of poor permeability or slow transport. Several approaches, like as nanotechnology, have been investigated to improve the distribution of drugs across the BBB [40]. Drug delivery by nanoparticles appears to be a potential option. Their targeting efficacy can be increased by conjugating them with certain ligands like aptamers, peptides, or antibodies. For example, adding TGN peptide and AS1411 aptamer to nanoparticles led to an increase in brain accumulation and BBB permeability [9]. Moreover, shuttle peptides made of d-amino acids have demonstrated effective and non-toxic BBB penetration. Transferrin receptor-targeted gold nanoparticles (TfR-targeted AuNPs) and venom-derived cyclic peptides have also shown promise for drug delivery to the brain [41]. Moreover, methods like ultrasound can momentarily enhance BBB permeability, which makes medication transit easier. By increasing BBB permeability without causing tissue injury, microbubble-enhanced diagnostic ultrasonography (MUES) has proven beneficial in the treatment of gliomas [42]. Even with these developments, there are still issues to be resolved, such as designing nanoparticles, increasing the effectiveness of carrier loading, and handling any possible harmful consequences. Furthermore, the intranasal route that circumvents the BBB by going through the olfactory pathway has shown promise as an effective means of delivering drugs to the brain [43].

In neurodegenerative diseases, there is considerable promise for both protective and degenerative effects from nanoparticles (Fig. **1**) that target the Central Nervous System (CNS). These minuscule transporters provide a potentially effective means of administering drugs that can protect neurons from harm and prolong their lifespan [44]. Nanoparticles protect neuroprotective drugs, antioxidants, or growth hormones against deterioration and guarantee their efficient transport to the central nervous system. Furthermore, the duration and effectiveness of therapeutic actions are further increased by their capacity to improve bioavailability through encapsulation. Nanoparticles can precisely target certain cell types or locations inside the Central Nervous System (CNS) by functionalizing them with ligands [45, 46]. Drug release from controlled-release formulations is optimized for maximum efficacy by responding to cues particular to the condition. Furthermore, gene therapy applications using nanoparticles seem promising as they can carry nucleic acids such as siRNA to modify gene expression linked to neurodegeneration and possibly change the progression of the disease [47].

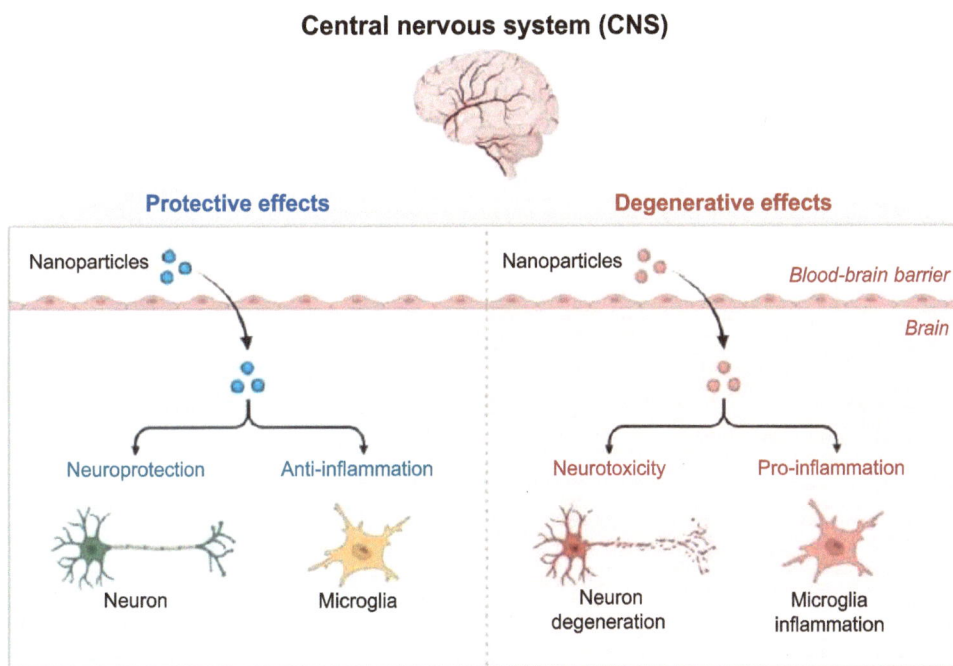

Fig. (1). Targeting CNS through nanoformulation.

Mechanism of Action of Nanoparticles

When it comes to delivering therapeutic compounds to the brain, nanoparticles can overcome the Blood-Brain Barrier (BBB) due to their distinct physiochemical characteristics. Important roles in this process are played by their small size, hydrophilicity, surface charge, and capacity to act as nanocarriers [48]. Electrostatic interactions increase the interaction between brain endothelial cells and nanoparticles. The BBB's negatively charged endothelial cells can bind to positively charged nanoparticles, facilitating their passage over the barrier.

Furthermore, the characteristics of nanoparticles are improved by their lipophilicity, which encourages adsorption and facilitates contact with the endothelium membrane [49]. Upon crossing the blood-brain barrier, nanoparticles can be taken up by endocytosis or transcytosis. Brain endothelial cells' low-density lipoprotein receptors help to make these events occur. The drugs are then released from their capsules onto the BBB's surface, where they diffuse into the brain parenchyma. Nanoparticles can improve medication delivery across the blood-brain barrier in several ways. A possible reason for enhanced membrane permeability is the solubilization of lipids in the membrane of endothelial cells [50]. Furthermore, efflux transporters, such as P-glycoprotein (P-gp), which normally block the flow of chemicals that are detrimental to the brain, may be inhibited by nanoparticles.

Organic Nano Formulation

To target the Central Nervous System (CNS), bridge the Blood-Brain Barrier (BBB), and treat neurological illnesses, polymeric nanoparticles and liposomes are potential drug delivery vehicles [51]. The capacity of polymeric nanoparticles to encapsulate and transport therapeutic compounds to the brain has been the subject of much research, especially about those composed of poly(lactide-c--glycolic) acid (PLGA) [52]. PLGA nanoparticles have demonstrated improved delivery of drugs to the brain in Alzheimer's disease and brain cancer, lowering oxidative stress, inflammation, and plaque burden. Similar to this, poly(ethylene imine) (PEI) based nanoparticles have shown promise for improved BBB passage and gene delivery in brain cancer treatment [53]. Nanoparticles of poly(allylamine) hydrochloride show promise as a treatment for neurological conditions as well as neuroprotective properties. The conjugation of chitosan with l-valine has demonstrated potential in the delivery of hydrophilic drugs for the treatment of Alzheimer's disease [54]. Compared to conventional formulations, penetrating amphiphilic polymer-lipid nanoparticles have successfully administered docetaxel for the treatment of brain metastases, with enhanced tumor growth suppression. Liposomes are synthetic spherical vesicles made of lipid bilayers that are frequently employed to transport medications to the central nervous system. To distribute several anticancer drugs, including methotrexate, 5-fluorouracil, paclitaxel, doxorubicin, and erlotinib, liposomal formulations have been used in brain cancer therapy [55]. Using poly(ethylene glycol) coating on liposomes to prolong circulation duration, transferrin for receptor targeting, and glucose-vitamin C complex for better accumulation at the target location are some techniques to promote BBB penetration. The liposomes modified with apolipoprotein E and transferrin have been used to treat Alzheimer's disease and neurodegenerative diseases, respectively. These liposomes have demonstrated improved blood-brain barrier penetration and effective drug administration in both *in vitro* and *in vivo* models [56].

Synthetic macromolecules known as dendrimers, which resemble trees, have shown promise as nanotechnology-based approaches to treating a variety of neurological disorders including crossing the Blood-Brain Barrier (BBB) [57]. Their encapsulating characteristics and fixed molecular weights make them perfect for targeted drug delivery. In particular, polyamidoamine dendrimers have been thoroughly researched for the therapy of brain disorders [58]. Glioblastoma treatment has demonstrated improved tumor penetration by the conjugation of glioma homing peptides and poly(ethylene glycol) to polyamidoamine dendrimers, whereas concanavalin, sialic acid, and glucosamine For the treatment of brain cancer, anchoring dendritic nanoconjugates have greatly improved the delivery of drugs to tumor locations [59].

Polyamidoamine dendrimers have encapsulated the anti-epileptic drug carbamazepine, with encouraging results in the treatment of Alzheimer's disease. In the treatment of ischemic stroke, they have also been used as delivery systems for drugs and to lessen blood clotting [60]. Moreover, by enabling medication distribution over the BBB and preventing neuroinflammation pathways, dendrimers have shown effective in treating cerebral palsy and hypothermic circulatory arrest. Oral administration of capsule-enhanced dendrimer formulations has demonstrated potential for the treatment of neuroinflammation [61].

Inorganic Nano-formulations

The term "inorganic-based nanoparticles" (NPs) refers to a broad class of particles that come in sizes varying from 1 to 100 nm and have different compositions and shapes. These NPs' distinct qualities and functions make them very promising for use in biological applications. Metallic NPs, including gold (AuNPs) and silver (AgNPs), are a well-known subgroup of inorganic-based NPs that have attracted a lot of interest because of their beneficial properties [62]. Due to their low toxicity, simple surface functionalization, and biocompatibility, AuNPs have been studied extensively and have potential in many biological areas. Functional groups like amines, phosphines, and thiols can be added to their surfaces to enable a variety of binding arrangements with biomolecules [63]. Similar to AgNPs, these compounds have broad-spectrum antibacterial and anticancer properties, but further research is needed to determine the safety of AgNPs' effects on neuroinflammation and brain development. Graphene, carbon nanotubes, and Carbon Dots (CDs) are examples of carbon-based nanomaterials that have shown promise in applications related to neurology [64]. Specifically, CDs have advantageous features including biocompatibility, compact size, and adjustable optical qualities. The Blood-Brain Barrier (BBB) may be crossed by CDs, as shown by several studies, which may allow for the targeted delivery of drugs to the brain. However, a careful analysis of CDs' safety profile is required [65]. In biomedical research, Magnetic Nanoparticles (MNPs)—most notably, iron oxide nanoparticles like Fe_3O_4 —offer special benefits including customizable form, high stability, and superparamagnetism. These characteristics make it easier to use them in medication delivery systems, which may have therapeutic benefits for brain disorders [66]. Through a variety of pathways, including transcytosis and magnetically mediated dragging, MNPs can cross the blood-brain barrier, providing new opportunities for targeted treatments and neuroprotective approaches. Another family of inorganic-based NPs with enzymatic activity is called nanozymes [67]. They resemble natural enzymes and are more stable and resilient to extreme environments. Functionalized nanozymes, based on boehmite NPs, have the potential to disassemble protein deposits linked to disorders like Huntington's and scavenge reactive oxygen species, therefore bridging the blood-brain barrier and reducing neurodegenerative processes [68].

Stimuli-Responsive Nanoparticles

Stimuli-responsive nanoparticles are useful for treating diseases including Parkinson's, Alzheimer's, and neuroinflammation because they are made to release drugs in response to pH, enzyme activity, or other local variables. With fewer systemic adverse effects, these methods enhance therapeutic efficacy by improving drug stability and retention in the nasal mucosa. A study optimized a mucoadhesive thermosensitive in situ gel for intranasal delivery of rivastigmine tartrate (RV) using pluronic F127 and Carbopol 934 to enhance brain targeting and bioavailability. The gel demonstrated improved nasal permeation (84%) and brain distribution (0.54%ID/g) compared to intranasal and intravenous RV solutions. Pharmacokinetic and biodistribution results highlighted its potential as an effective system for transnasal drug delivery to the brain [69].

One more study developed and optimized an ion-activated in-situ gel for the intranasal delivery of rizatriptan benzoate to enhance brain targeting. The formulation showed improved gel strength, mucoadhesive properties, and drug release compared to solutions, demonstrating its potential as an effective alternative for brain delivery *via* the nasal route [70].

Neurotoxicity of Nano-formulations

When it comes to treating neuroinflammation in particular, nanomaterials provide viable paths for achieving the vital therapeutic objective of counteracting inflammation. However, given that many atoms might become extremely reactive due to the nanoparticles' small size, there are still worries about their possible toxicity, particularly in the context of neuroinflammation treatment [71]. Thus, it becomes critical to comprehend and reduce neurotoxicity. Any detrimental effect that physical or chemical stimuli can have on the chemistry, structure, or function of the nervous system is referred to as neurotoxicity [72]. The main processes include the release of cytokines, which causes neuroinflammation, the overproduction of reactive oxygen species, which results in oxidative stress, and the dysregulation of apoptosis, which causes neuronal death. Numerous properties of nanoparticles, such as size and surface area, have a complex relationship with neurotoxicity [73]. Researchers have carefully examined widely used nanoparticles for possible neurotoxic consequences. For example, astrogliosis, which is characterized by an increase in the number and size of astrocytes as well as cognitive deficiencies such as attention and memory problems, has been linked to gold nanoparticles (AuNPs). Astrogliosis is associated with hypoxia, ischemia, and seizures and frequently coexists with brain illnesses such as Alzheimer's. High dosages of anatase TiO_2 nanoparticles increase brain and plasma levels of IL-6, suggesting a possible activation of neuroinflammation [74]. Exposure to iron oxide nanoparticles (IONPs) can cause oxidative damage in particular brain areas, interfere with synaptic transmission and nerve conduction, and encourage immune cell infiltration, neural inflammation, and death. Polysorbate 80-modified chitosan nanoparticles and drug-free liposomes have demonstrated neuropathological alterations such as necrosis, neuroinflammation, neuronal apoptosis, and elevated oxidative stress. Compared to their organic counterparts, inorganic nanoparticles often show more frequent and severe toxicity [75]. It is interesting to note that certain nanoparticles can cause neuroinflammation, even though the majority show anti-neuroinflammatory effects either on their own or when combined with anti-inflammatory drugs [76].

AI Integration in Nanocarrier Design for Neurodegenerative Disorders

In the optimization of nanocarrier design for Neurodegenerative Diseases (NDs), artificial intelligence (AI) (Fig. **2**) has become an essential tool, tackling important issues including the efficient delivery of drugs and Blood-Brain Barrier (BBB) crossing. Machine Learning (ML), one of the fundamental AI techniques, is essential since it analyzes intricate datasets to determine the ideal nanocarrier characteristics. To provide accurate therapeutic administration, it is especially useful in predicting drug-nanocarrier interactions and optimizing parameters like drug loading and release patterns. This is furthered by Deep Learning (DL), which models molecular biological interactions. This makes it possible to develop nanocarriers that are specifically made to penetrate the blood-brain barrier, frequently by mimicking lipid-based or polymeric formulations that improve the distribution of drugs to specific areas [77]. Large-scale datasets help both ML and

DL techniques to enhance treatment results and their prediction accuracy. Computational modeling predicts the biodistribution and pharmacokinetics of nanocarriers, enabling precise targeting of diseased brain regions by simulating the behavior of different formulations *in vivo*. Big Data analytics combines various patient-specific datasets, such as genetic, proteomic, and metabolomic profiles, allowing for personalized nanocarrier designs. This approach guarantees that the therapeutic strategy is tailored to an individual's unique disease profile, potentially increasing treatment efficacy. There are several benefits of using AI in the development of nanocarriers [78]. AI-driven models increase drug delivery systems' specificity, guaranteeing better targeting of impacted brain areas and reducing side effects. Predictive models also aid in anticipating treatment results, which speeds up the drug development process and improves clinical trial design. Crucially, by reducing trial-and-error experimentation and the corresponding resource consumption, these methods reduce expenses. These technologies not only improve the effectiveness of nanocarriers but also open the door for the creation of individualized treatment plans by utilizing AI's capacity to handle and evaluate massive information. Although there are still issues with data quality, interpretability, and ethics, the combination of AI and nanotechnology has enormous potential to transform ND therapy, provide hope for better patient outcomes, and push the boundaries of precision medicine [79].

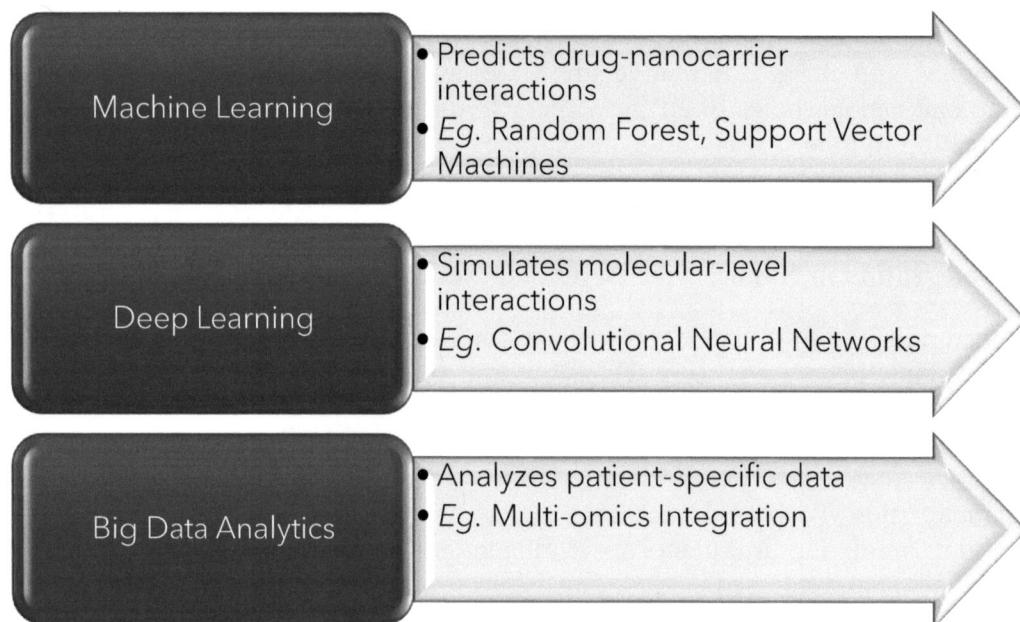

Machine Learning
- Predicts drug-nanocarrier interactions
- *Eg.* Random Forest, Support Vector Machines

Deep Learning
- Simulates molecular-level interactions
- *Eg.* Convolutional Neural Networks

Big Data Analytics
- Analyzes patient-specific data
- *Eg.* Multi-omics Integration

Fig. (2). AI integration in nanocarrier design for neurodegenerative disorders.

CONCLUSION

While the BBB is essential for preserving CNS homeostasis, it also poses a significant barrier to the administration of drugs for neurological conditions. Innovative ways to circumvent or alter the blood-brain barrier are made possible by nanotechnology, which also improves the delivery of drugs to the brain. Targeting neurodegenerative processes and improving drug delivery are two areas in which polymeric nanoparticles, liposomes, dendrimers, and inorganic nanoparticles have demonstrated substantial promise. Although nanoparticles have many interesting uses, their neurotoxicity is still a concern that requires careful safety assessments. Moreover, natural compounds contained in nanoparticles offer a cutting-edge method for improving bioavailability and therapeutic efficacy in treating neurological disorders. Prolonged multidisciplinary study on nanotechnology and BBB modulation has the potential to improve patient outcomes by progressing the identification and management of brain diseases. Furthermore, integrating AI and nanotechnology offers transformative potential for precision-targeted, cost-effective treatments for neurodegenerative disorders.

REFERENCES

[1]   Wu, D.; Chen, Q.; Chen, X.; Han, F.; Chen, Z.; Wang, Y. The blood–brain barrier: Structure, regulation and drug delivery. *Signal Transduct. Target. Ther.,* **2023,** *8*(1), 217.
[http://dx.doi.org/10.1038/s41392-023-01481-w] [PMID: 37231000]

[2]   Dotiwala, A.K.; McCausland, C.; Samra, N.S. *Anatomy, Head and Neck, Blood Brain Barrier*; StatPearls Publishing, **2020.**

[3]   Miller, L.M.; Gal, A. *Cardiovascular System and Lymphatic Vessels*; Pathol. Basis Vet. Dis. Expert Consult, **2017,** pp. 561-616.e1.
[http://dx.doi.org/10.1016/B978-0-323-35775-3.00010-2]

[4]   Alberts, B; Johnson, A; Lewis, J; Raff, M; Roberts, K; Walter, P. Blood vessels and endothelial cells. *Mol Biol Cell,* **2007.**

[5]   Mironov, A.A.; Mironov, A.; Sanavio, B.; Krol, S.; Beznoussenko, G.V. Intracellular Membrane Transport in Vascular Endothelial Cells. *Int. J. Mol. Sci.,* **2023,** *24*(6), 5791.
[http://dx.doi.org/10.3390/ijms24065791] [PMID: 36982865]

[6]   Kumar, A.; Tan, A.; Wong, J.; Spagnoli, J.C.; Lam, J.; Blevins, B.D.; G, N.; Thorne, L.; Ashkan, K.; Xie, J.; Liu, H. Nanotechnology for Neuroscience: Promising Approaches for Diagnostics, Therapeutics and Brain Activity Mapping. *Adv. Funct. Mater.,* **2017,** *27*(39), 1700489.
[http://dx.doi.org/10.1002/adfm.201700489] [PMID: 30853878]

[7]   Markowicz-Piasecka, M.; Darłak, P.; Markiewicz, A.; Sikora, J.; Kumar Adla, S.; Bagina, S.; Huttunen, K.M. Current approaches to facilitate improved drug delivery to the central nervous system. *Eur. J. Pharm. Biopharm.,* **2022,** *181*, 249-262.
[http://dx.doi.org/10.1016/j.ejpb.2022.11.003] [PMID: 36372271]

[8]   Kyriakides, T.R.; Raj, A.; Tseng, T.H.; Xiao, H.; Nguyen, R.; Mohammed, F.S.; Halder, S.; Xu, M.; Wu, M.J.; Bao, S.; Sheu, W.C. Biocompatibility of nanomaterials and their immunological properties. *Biomed. Mater.,* **2021,** *16*(4), 042005.
[http://dx.doi.org/10.1088/1748-605X/abe5fa] [PMID: 33578402]

[9]   Pinheiro, R.G.R.; Coutinho, A.J.; Pinheiro, M.; Neves, A.R. Nanoparticles for targeted brain drug delivery: What do we know? *Int. J. Mol. Sci.,* **2021**, *22*(21), 11654.
[http://dx.doi.org/10.3390/ijms222111654] [PMID: 34769082]

[10]   Adir, O.; Poley, M.; Chen, G.; Froim, S.; Krinsky, N.; Shklover, J.; Shainsky-Roitman, J.; Lammers, T.; Schroeder, A. Integrating Artificial Intelligence and Nanotechnology for Precision Cancer Medicine. *Adv. Mater.,* **2020**, *32*(13), 1901989.
[http://dx.doi.org/10.1002/adma.201901989] [PMID: 31286573]

[11]   Serov, N.; Vinogradov, V. Artificial intelligence to bring nanomedicine to life. *Adv. Drug Deliv. Rev.,* **2022**, *184*, 114194.
[http://dx.doi.org/10.1016/j.addr.2022.114194] [PMID: 35283223]

[12]   Scalise, A.A.; Kakogiannos, N.; Zanardi, F.; Iannelli, F.; Giannotta, M. The blood–brain and gut–vascular barriers: from the perspective of claudins. *Tissue Barriers,* **2021**, *9*(3), 1926190.
[http://dx.doi.org/10.1080/21688370.2021.1926190] [PMID: 34152937]

[13]   Knox, E.G.; Aburto, M.R.; Clarke, G.; Cryan, J.F.; O'Driscoll, C.M. The blood-brain barrier in aging and neurodegeneration. *Mol. Psychiatry,* **2022**, *27*(6), 2659-2673.
[http://dx.doi.org/10.1038/s41380-022-01511-z] [PMID: 35361905]

[14]   Jagtiani, E.; Yeolekar, M.; Naik, S.; Patravale, V. *In vitro* blood brain barrier models: An overview. *J. Control. Release,* **2022**, *343*, 13-30.
[http://dx.doi.org/10.1016/j.jconrel.2022.01.011] [PMID: 35026351]

[15]   Pandit, R.; Chen, L.; Götz, J. The blood-brain barrier: Physiology and strategies for drug delivery. *Adv. Drug Deliv. Rev.,* **2020**, *165-166*, 1-14.
[http://dx.doi.org/10.1016/j.addr.2019.11.009] [PMID: 31790711]

[16]   Archie, S.R.; Al Shoyaib, A.; Cucullo, L. Blood-brain barrier dysfunction in cns disorders and putative therapeutic targets: An overview. *Pharmaceutics,* **2021**, *13*(11), 1779.
[http://dx.doi.org/10.3390/pharmaceutics13111779] [PMID: 34834200]

[17]   Gosselet, F.; Loiola, R.A.; Roig, A.; Rosell, A.; Culot, M. Central nervous system delivery of molecules across the blood-brain barrier. *Neurochem. Int.,* **2021**, *144*, 104952.
[http://dx.doi.org/10.1016/j.neuint.2020.104952] [PMID: 33400964]

[18]   Olufunmilayo, E.O.; Gerke-Duncan, M.B.; Holsinger, R.M.D. Oxidative Stress and Antioxidants in Neurodegenerative Disorders. *Antioxidants,* **2023**, *12*(2), 517.
[http://dx.doi.org/10.3390/antiox12020517] [PMID: 36830075]

[19]   Gawdi, R.; Emmady, P.D. *Physiology, Blood Brain Barrier*; StatPearls Publishing, **2020**.

[20]   Kadry, H.; Noorani, B.; Cucullo, L. A blood–brain barrier overview on structure, function, impairment, and biomarkers of integrity. *Fluids Barriers CNS,* **2020**, *17*(1), 69.
[http://dx.doi.org/10.1186/s12987-020-00230-3] [PMID: 33208141]

[21]   Vaillant, A.A.J.; Ahmad, F. *Leukocyte Adhesion Deficiency*; StatPearls, **2023**.

[22]   Gullotta, G.S.; Costantino, G.; Sortino, M.A.; Spampinato, S.F. Microglia and the Blood–Brain Barrier: An External Player in Acute and Chronic Neuroinflammatory Conditions. *Int. J. Mol. Sci.,* **2023**, *24*(11), 9144.
[http://dx.doi.org/10.3390/ijms24119144] [PMID: 37298096]

[23]   Haqqani, A.S.; Bélanger, K.; Stanimirovic, D.B. Receptor-mediated transcytosis for brain delivery of therapeutics: receptor classes and criteria. *Frontiers in Drug Delivery,* **2024**, *4*, 1360302.
[http://dx.doi.org/10.3389/fddev.2024.1360302]

[24]   Parodi, A.; Rudzińska, M.; Deviatkin, A.; Soond, S.; Baldin, A.; Zamyatnin, A., Jr Established and emerging strategies for drug delivery across the blood-brain barrier in brain cancer. *Pharmaceutics,* **2019**, *11*(5), 245.
[http://dx.doi.org/10.3390/pharmaceutics11050245] [PMID: 31137689]

[25]   Pandey, P.K.; Sharma, A.K.; Gupta, U. Blood brain barrier: An overview on strategies in drug delivery, realistic *in vitro* modeling and *in vivo* live tracking. *Tissue Barriers,* **2016**, *4*(1), e1129476.
[http://dx.doi.org/10.1080/21688370.2015.1129476] [PMID: 27141418]

[26]   Zhao, Y.; Gan, L.; Ren, L.; Lin, Y.; Ma, C.; Lin, X. Factors influencing the blood-brain barrier permeability. *Brain Res.,* **1788**, *2022*
[http://dx.doi.org/10.1016/j.brainres.2022.147937] [PMID: 35568085]

[27]   Lin, A.; Peiris, N.J.; Dhaliwal, H.; Hakim, M.; Li, W.; Ganesh, S.; Ramaswamy, Y.; Patel, S.; Misra, A. Mural cells: Potential therapeutic targets to bridge cardiovascular disease and neurodegeneration. *Cells,* **2021**, *10*(3), 593.
[http://dx.doi.org/10.3390/cells10030593] [PMID: 33800271]

[28]   Uemura, M.T.; Maki, T.; Ihara, M.; Lee, V.M.Y.; Trojanowski, J.Q. Brain Microvascular Pericytes in Vascular Cognitive Impairment and Dementia. *Front. Aging Neurosci.,* **2020**, *12*, 80.
[http://dx.doi.org/10.3389/fnagi.2020.00080] [PMID: 32317958]

[29]   Bohannon, D.G.; Long, D.; Kim, W.K. Understanding the heterogeneity of human pericyte subsets in blood–brain barrier homeostasis and neurological diseases. *Cells,* **2021**, *10*(4), 890.
[http://dx.doi.org/10.3390/cells10040890] [PMID: 33919664]

[30]   Norris, G.T.; Kipnis, J. Immune cells and CNS physiology: Microglia and beyond. *J. Exp. Med.,* **2019**, *216*(1), 60-70.
[http://dx.doi.org/10.1084/jem.20180199] [PMID: 30504438]

[31]   Zhou, B.; Zuo, Y.X.; Jiang, R.T. Astrocyte morphology: Diversity, plasticity, and role in neurological diseases. *CNS Neurosci. Ther.,* **2019**, *25*(6), 665-673.
[http://dx.doi.org/10.1111/cns.13123] [PMID: 30929313]

[32]   Han, L. Modulation of the blood–brain barrier for drug delivery to brain. *Pharmaceutics,* **2021**, *13*(12), 2024.
[http://dx.doi.org/10.3390/pharmaceutics13122024] [PMID: 34959306]

[33]   Jahncke, J.N.; Wright, K.M. The many roles of dystroglycan in nervous system development and function. *Dev. Dyn.,* **2023**, *252*(1), 61-80.
[http://dx.doi.org/10.1002/dvdy.516] [PMID: 35770940]

[34]   Liu, Z.; Zhou, T.; Ziegler, A.C.; Dimitrion, P.; Zuo, L. Oxidative Stress in Neurodegenerative Diseases: From Molecular Mechanisms to Clinical Applications. *Oxid. Med. Cell. Longev.,* **2017**, *2017*(1), 2525967.
[http://dx.doi.org/10.1155/2017/2525967] [PMID: 28785371]

[35]   Haque, R.; Alam, K.; Gow, J.; Neville, C. Changes in the prevalence of dementia in Australia and its association with geographic remoteness. *PLoS One,* **2023**, *18*(8), e0289505.
[http://dx.doi.org/10.1371/journal.pone.0289505] [PMID: 37531396]

[36]   García, J.C.; Bustos, R.H. The genetic diagnosis of neurodegenerative diseases and therapeutic perspectives. *Brain Sci.,* **2018**, *8*(12), 222.
[http://dx.doi.org/10.3390/brainsci8120222] [PMID: 30551598]

[37]   Adam, H; Gopinath, SCB; Md Arshad, MK; Adam, T; Parmin, NA; Husein, I An update on pathogenesis and clinical scenario for Parkinson's disease: diagnosis and treatment. *3 Biotech,* **2023**, 13.
[http://dx.doi.org/10.1007/s13205-023-03553-8]

[38]   Barichello, T.; Collodel, A.; Hasbun, R.; Morales, R. An Overview of the Blood-Brain Barrier. *Neuromethods,* **2019**, *142*, 1-8.
[http://dx.doi.org/10.1007/978-1-4939-8946-1_1]

[39]   Pardridge, W.M. A historical review of brain drug delivery. *Pharmaceutics,* **2022**, *14*(6), 1283.
[http://dx.doi.org/10.3390/pharmaceutics14061283] [PMID: 35745855]

[40] Formica, M.L.; Real, D.A.; Picchio, M.L.; Catlin, E.; Donnelly, R.F.; Paredes, A.J. On a highway to the brain: A review on nose-to-brain drug delivery using nanoparticles. *Appl. Mater. Today,* **2022**, *29*, 101631.
[http://dx.doi.org/10.1016/j.apmt.2022.101631]

[41] Sánchez-Navarro, M.; Giralt, E. Peptide Shuttles for Blood–Brain Barrier Drug Delivery. *Pharmaceutics,* **2022**, *14*(9), 1874.
[http://dx.doi.org/10.3390/pharmaceutics14091874] [PMID: 36145622]

[42] Wu, S.K.; Tsai, C.L.; Huang, Y.; Hynynen, K. Focused ultrasound and microbubbles-mediated drug delivery to brain tumor. *Pharmaceutics,* **2020**, *13*(1), 15.
[http://dx.doi.org/10.3390/pharmaceutics13010015] [PMID: 33374205]

[43] Chehelgerdi, M.; Chehelgerdi, M.; Allela, O.Q.B.; Pecho, R.D.C.; Jayasankar, N.; Rao, D.P.; Thamaraikani, T.; Vasanthan, M.; Viktor, P.; Lakshmaiya, N.; Saadh, M.J.; Amajd, A.; Abo-Zaid, M.A.; Castillo-Acobo, R.Y.; Ismail, A.H.; Amin, A.H.; Akhavan-Sigari, R. Progressing nanotechnology to improve targeted cancer treatment: overcoming hurdles in its clinical implementation. *Mol. Cancer,* **2023**, *22*(1), 169.
[http://dx.doi.org/10.1186/s12943-023-01865-0] [PMID: 37814270]

[44] Rhaman, M.M.; Islam, M.R.; Akash, S.; Mim, M.; Noor alam, M.; Nepovimova, E.; Valis, M.; Kuca, K.; Sharma, R. Exploring the role of nanomedicines for the therapeutic approach of central nervous system dysfunction: At a glance. *Front. Cell Dev. Biol.,* **2022**, *10*, 989471.
[http://dx.doi.org/10.3389/fcell.2022.989471] [PMID: 36120565]

[45] Yetisgin, A.A.; Cetinel, S.; Zuvin, M.; Kosar, A.; Kutlu, O. Therapeutic nanoparticles and their targeted delivery applications. *Molecules,* **2020**, *25*(9), 2193.
[http://dx.doi.org/10.3390/molecules25092193] [PMID: 32397080]

[46] Scarpa, E.; Cascione, M.; Griego, A.; Pellegrino, P.; Moschetti, G.; De Matteis, V. Gold and silver nanoparticles in Alzheimer's and Parkinson's diagnostics and treatments. *Ibrain,* **2023**, *9*(3), 298-315.
[http://dx.doi.org/10.1002/ibra.12126] [PMID: 37786760]

[47] Finocchio, L.; Zeppieri, M.; Gabai, A.; Toneatto, G.; Spadea, L.; Salati, C. Recent Developments in Gene Therapy for Neovascular Age-Related Macular Degeneration: A Review. *Biomedicines,* **2023**, *11*(12), 3221.
[http://dx.doi.org/10.3390/biomedicines11123221] [PMID: 38137442]

[48] Teleanu, D.M.; Chircov, C.; Grumezescu, A.M.; Volceanov, A.; Teleanu, R.I. Blood-brain delivery methods using nanotechnology. *Pharmaceutics,* **2018**, *10*(4), 269.
[http://dx.doi.org/10.3390/pharmaceutics10040269] [PMID: 30544966]

[49] Hersh, A.M.; Alomari, S.; Tyler, B.M. Crossing the Blood-Brain Barrier: Advances in Nanoparticle Technology for Drug Delivery in Neuro-Oncology. *Int. J. Mol. Sci.,* **2022**, *23*(8), 4153.
[http://dx.doi.org/10.3390/ijms23084153] [PMID: 35456971]

[50] Bellettato, CM; Scarpa, M Possible strategies to cross the blood-brain barrier. *Ital J Pediatr,* **2018**, 44.
[http://dx.doi.org/10.1186/s13052-018-0563-0]

[51] Annu, S.A.; Sartaj, A.; Qamar, Z.; Md, S.; Alhakamy, N.A.; Baboota, S.; Ali, J. An Insight to Brain Targeting Utilizing Polymeric Nanoparticles: Effective Treatment Modalities for Neurological Disorders and Brain Tumor. *Front. Bioeng. Biotechnol.,* **2022**, *10*, 788128.
[http://dx.doi.org/10.3389/fbioe.2022.788128] [PMID: 35186901]

[52] Begines, B.; Ortiz, T.; Pérez-Aranda, M.; Martínez, G.; Merinero, M.; Argüelles-Arias, F.; Alcudia, A. Polymeric nanoparticles for drug delivery: Recent developments and future prospects. *Nanomaterials (Basel),* **2020**, *10*(7), 1403.
[http://dx.doi.org/10.3390/nano10071403] [PMID: 32707641]

[53] Cunha, A.; Gaubert, A.; Latxague, L.; Dehay, B. PLGA-based nanoparticles for neuroprotective drug delivery in neurodegenerative diseases. *Pharmaceutics,* **2021**, *13*(7), 1042.

[http://dx.doi.org/10.3390/pharmaceutics13071042] [PMID: 34371733]

[54]   Poudel, P.; Park, S. Recent Advances in the Treatment of Alzheimer's Disease Using Nanoparticle-Based Drug Delivery Systems. *Pharmaceutics,* **2022,** *14*(4), 835.
[http://dx.doi.org/10.3390/pharmaceutics14040835] [PMID: 35456671]

[55]   Olusanya, T.; Haj Ahmad, R.; Ibegbu, D.; Smith, J.; Elkordy, A. Liposomal drug delivery systems and anticancer drugs. *Molecules,* **2018,** *23*(4), 907.
[http://dx.doi.org/10.3390/molecules23040907] [PMID: 29662019]

[56]   Pandian, S.R.K.; Vijayakumar, K.K.; Murugesan, S.; Kunjiappan, S. Liposomes: An emerging carrier for targeting Alzheimer's and Parkinson's diseases. *Heliyon,* **2022,** *8*(6), e09575.
[http://dx.doi.org/10.1016/j.heliyon.2022.e09575] [PMID: 35706935]

[57]   Zhu, Y.; Liu, C.; Pang, Z. Dendrimer-based drug delivery systems for brain targeting. *Biomolecules,* **2019,** *9*(12), 790.
[http://dx.doi.org/10.3390/biom9120790] [PMID: 31783573]

[58]   Florendo, M.; Figacz, A.; Srinageshwar, B.; Sharma, A.; Swanson, D.; Dunbar, G.L.; Rossignol, J. Use of polyamidoamine dendrimers in brain diseases. *Molecules,* **2018,** *23*(9), 2238.
[http://dx.doi.org/10.3390/molecules23092238] [PMID: 30177605]

[59]   Kaurav, M.; Ruhi, S.; Al-Goshae, H.A.; Jeppu, A.K.; Ramachandran, D.; Sahu, R.K.; Sarkar, A.K.; Khan, J.; Ashif Ikbal, A.M. Dendrimer: An update on recent developments and future opportunities for the brain tumors diagnosis and treatment. *Front. Pharmacol.,* **2023,** *14*, 1159131.
[http://dx.doi.org/10.3389/fphar.2023.1159131] [PMID: 37006997]

[60]   Igartúa, D.E.; Martinez, C.S.; Temprana, C.F.; Alonso, S.V.; Prieto, M.J. PAMAM dendrimers as a carbamazepine delivery system for neurodegenerative diseases: A biophysical and nanotoxicological characterization. *Int. J. Pharm.,* **2018,** *544*(1), 191-202.
[http://dx.doi.org/10.1016/j.ijpharm.2018.04.032] [PMID: 29678547]

[61]   Zhang, F.; Zhang, Z.; Alt, J.; Kambhampati, S.P.; Sharma, A.; Singh, S.; Nance, E.; Thomas, A.G.; Rojas, C.; Rais, R.; Slusher, B.S.; Kannan, R.M.; Kannan, S. Dendrimer-enabled targeted delivery attenuates glutamate excitotoxicity and improves motor function in a rabbit model of cerebral palsy. *J. Control. Release,* **2023,** *358*, 27-42.
[http://dx.doi.org/10.1016/j.jconrel.2023.04.017] [PMID: 37054778]

[62]   Kanakari, E.; Dendrinou-Samara, C. Fighting Phytopathogens with Engineered Inorganic-Based Nanoparticles. *Materials (Basel),* **2023,** *16*(6), 2388.
[http://dx.doi.org/10.3390/ma16062388] [PMID: 36984268]

[63]   Kumalasari, M.R.; Alfanaar, R.; Andreani, A.S. Gold nanoparticles (AuNPs): A versatile material for biosensor application. *Talanta Open,* **2024,** *9*, 100327.
[http://dx.doi.org/10.1016/j.talo.2024.100327]

[64]   Pandey, R.R.; Chusuei, C.C. Carbon nanotubes, graphene, and carbon dots as electrochemical biosensing composites. *Molecules,* **2021,** *26*(21), 6674.
[http://dx.doi.org/10.3390/molecules26216674] [PMID: 34771082]

[65]   Sun, Y.; Du, L.; Yang, M.; Li, Q.; Jia, X.; Li, Q.; Zhu, L.; Zhang, Y.; Liu, Y.; Liu, S. Brain-targeted drug delivery assisted by physical techniques and its potential applications in traditional Chinese medicine. *Journal of Traditional Chinese Medical Sciences,* **2021,** *8*(3), 186-197.
[http://dx.doi.org/10.1016/j.jtcms.2021.07.003]

[66]   Ganapathe, L.S.; Mohamed, M.A.; Mohamad Yunus, R.; Berhanuddin, D.D. Magnetite (Fe$_3$O$_4$) nanoparticles in biomedical application: From synthesis to surface functionalisation. *Magnetochemistry,* **2020,** *6*(4), 68.
[http://dx.doi.org/10.3390/magnetochemistry6040068]

[67]   Zhao, Q.; Du, W.; Zhou, L.; Wu, J.; Zhang, X.; Wei, X.; Wang, S.; Huang, Y.; Li, Y. Transferrin-Enabled Blood–Brain Barrier Crossing Manganese-Based Nanozyme for Rebalancing the Reactive

Oxygen Species Level in Ischemic Stroke. *Pharmaceutics,* **2022,** *14*(6), 1122.
[http://dx.doi.org/10.3390/pharmaceutics14061122] [PMID: 35745695]

[68] Zhang, Y.; Zhang, L.; Wang, M.; Li, P. The applications of nanozymes in neurological diseases: From mechanism to design. *Theranostics,* **2023,** *13*(8), 2492-2514.
[http://dx.doi.org/10.7150/thno.83370] [PMID: 37215578]

[69] Abouhussein, D.M.N.; Khattab, A.; Bayoumi, N.A.; Mahmoud, A.F.; Sakr, T.M. Brain targeted rivastigmine mucoadhesive thermosensitive In situ gel: Optimization, *in vitro* evaluation, radiolabeling, *in vivo* pharmacokinetics and biodistribution. *J. Drug Deliv. Sci. Technol.,* **2018,** *43*, 129-140.
[http://dx.doi.org/10.1016/j.jddst.2017.09.021]

[70] Thakkar, J.H.; Prajapati, S.T. Formulation development and characterization of in-situ gel of Rizatriptan Benzoate for intranasal delivery. *J. Drug Deliv. Ther.,* **2021,** *11*(1-s), 1-6.
[http://dx.doi.org/10.22270/jddt.v11i1-s.4685]

[71] Cerqueira, S.R.; Ayad, N.G.; Lee, J.K. Neuroinflammation Treatment *via* Targeted Delivery of Nanoparticles. *Front. Cell. Neurosci.,* **2020,** *14*, 576037.
[http://dx.doi.org/10.3389/fncel.2020.576037] [PMID: 33192321]

[72] Serafini, M.M.; Sepehri, S.; Midali, M.; Stinckens, M.; Biesiekierska, M.; Wolniakowska, A.; Gatzios, A.; Rundén-Pran, E.; Reszka, E.; Marinovich, M.; Vanhaecke, T.; Roszak, J.; Viviani, B.; SenGupta, T. Recent advances and current challenges of new approach methodologies in developmental and adult neurotoxicity testing. *Arch. Toxicol.,* **2024,** *98*(5), 1271-1295.
[http://dx.doi.org/10.1007/s00204-024-03703-8] [PMID: 38480536]

[73] Huang, X.; He, D.; Pan, Z.; Luo, G.; Deng, J. Reactive-oxygen-species-scavenging nanomaterials for resolving inflammation. *Mater. Today Bio,* **2021,** *11*, 100124.
[http://dx.doi.org/10.1016/j.mtbio.2021.100124] [PMID: 34458716]

[74] Hambali, A.; Kumar, J.; Hashim, N.F.M.; Maniam, S.; Mehat, M.Z.; Cheema, M.S.; Mustapha, M.; Adenan, M.I.; Stanslas, J.; Hamid, H.A. Hypoxia-Induced Neuroinflammation in Alzheimer's Disease: Potential Neuroprotective Effects of *Centella asiatica. Front. Physiol.,* **2021,** *12*, 712317.
[http://dx.doi.org/10.3389/fphys.2021.712317] [PMID: 34721056]

[75] Ngowi, E.E.; Wang, Y.Z.; Qian, L.; Helmy, Y.A.S.H.; Anyomi, B.; Li, T.; Zheng, M.; Jiang, E.S.; Duan, S.F.; Wei, J.S.; Wu, D.D.; Ji, X.Y. The Application of Nanotechnology for the Diagnosis and Treatment of Brain Diseases and Disorders. *Front. Bioeng. Biotechnol.,* **2021,** *9*, 629832.
[http://dx.doi.org/10.3389/fbioe.2021.629832] [PMID: 33738278]

[76] Zhu, F.D.; Hu, Y.J.; Yu, L.; Zhou, X.G.; Wu, J.M.; Tang, Y.; Qin, D.L.; Fan, Q.Z.; Wu, A.G. Nanoparticles: A Hope for the Treatment of Inflammation in CNS. *Front. Pharmacol.,* **2021,** *12*, 683935.
[http://dx.doi.org/10.3389/fphar.2021.683935] [PMID: 34122112]

[77] Belić, M.; Bobić, V.; Badža, M.; Šolaja, N.; Đurić-Jovičić, M.; Kostić, V.S. Artificial intelligence for assisting diagnostics and assessment of Parkinson's disease—A review. *Clin. Neurol. Neurosurg.,* **2019,** *184*, 105442.
[http://dx.doi.org/10.1016/j.clineuro.2019.105442] [PMID: 31351213]

[78] Khastehband, S.; Ghasempour Dabbaghi, K.; Khosravirad, Z.; Jamalnia, S.; GhorbaniNia, R.; Mahmoudikohani, F.; Zakeri, H. The Use of Artificial Intelligence in the Management of Neurodegenerative Disorders; Focus on Alzheimer's Disease. *Galen Med. J.,* **2023,** *12*, e3061.
[http://dx.doi.org/10.31661/gmj.v12i3061] [PMID: 38827644]

[79] Dhankhar, S.; Mujwar, S.; Garg, N.; Chauhan, S.; Saini, M.; Sharma, P.; Kumar, S.; Kumar Sharma, S.; Kamal, M.A.; Rani, N. Artificial Intelligence in The Management of Neurodegenerative Disorders. *CNS Neurol. Disord. Drug Targets,* **2024,** *23*(8), 931-940.
[http://dx.doi.org/10.2174/0118715273266095231009092603] [PMID: 37861051]

CHAPTER 6

Clinical Strategies, Management, and Challenges in Targeting CNS through Bioactive Compounds

Abstract: The Central Nervous System (CNS) plays a pivotal role in regulating essential physiological processes, highlighting the critical need for effective therapeutic interventions targeting CNS disorders. Bioactive compounds, including small molecules and peptides, offer promising avenues for CNS drug discovery and development by modulating specific molecular targets implicated in neurological pathophysiology. However, the Blood-Brain Barrier (BBB) poses a formidable obstacle to CNS drug delivery, necessitating innovative approaches such as nanoparticle delivery systems, viral vectors, and alternative delivery methods like convection-enhanced delivery. This chapter comprehensively explores clinical strategies, management approaches, and challenges associated with targeting the CNS through bioactive compounds. It examines pharmacokinetic and pharmacodynamic considerations relevant to CNS drug delivery, highlighting innovative technologies and formulation approaches aimed at enhancing therapeutic efficacy. Despite significant progress, formidable challenges persist, including the complexity of CNS biology, the risk of off-target effects, and stringent regulatory requirements. Future research directions encompass developing better preclinical models, exploring multi-target approaches, and implementing early intervention strategies supported by advanced biomarkers to address the global burden of neurological disorders effectively.

Keywords: Animal models, Bioactive compound, CNS, Management, Personalized medicine.

INTRODUCTION

The Central Nervous System (CNS) serves as the command center of the human body, comprising the brain and spinal cord, and orchestrating a myriad of physiological processes essential for life. The intricate network of neurons and neurotransmitters within the CNS regulates sensory perception, motor function, cognition, and emotion, underscoring its indispensable role in human health and well-being. Consequently, dysfunction or pathology within the CNS can manifest as a diverse array of neurological disorders, ranging from neurodegenerative diseases like Alzheimer's and Parkinson's to psychiatric conditions such as depression and schizophrenia [1]. Given the profound impact of CNS disorders on

Shivendra Mani Tripathi, Sudhanshu Mishra, Rishabha Malviya & Smriti Ojha

individuals and society at large, there exists a pressing need for effective therapeutic interventions that can target the underlying mechanisms of these conditions. Bioactive compounds, including small molecules, peptides, and biologics, represent a promising avenue for CNS drug discovery and development due to their ability to modulate specific molecular targets implicated in neurological pathophysiology [2]. The scope of this chapter encompasses a comprehensive exploration of the clinical strategies, management approaches, and challenges associated with targeting the CNS through bioactive compounds. By elucidating the complex interplay between drug molecules and the CNS environment, this chapter aims to provide a nuanced understanding of the opportunities and obstacles inherent in this endeavor. In subsequent sections, we will examine in detail the pharmacokinetic and pharmacodynamic considerations relevant to CNS drug delivery, with a particular focus on overcoming the Blood-Brain Barrier (BBB) a formidable obstacle that limits the entry of therapeutics into the brain parenchyma [3]. The case studies presented in the discussion illustrate the treatment outcomes for various CNS illnesses, with several studies indicating enduring improvements over time. Patients with Parkinson's Disease (PD) who participated in exercise regimens, for instance, had sustained gains in strength, gait, and range of motion. The administration of Riluzole to patients with Amyotrophic Lateral Sclerosis (ALS) also resulted in stable functional scores over time, suggesting a good response even though the disease is progressing. Myasthenia gravis follow-up data showed improved quality-of-life ratings and successful self-management after discharge. Following intense physiotherapy sessions, long-term results from therapies such as intravenous immunoglobulin therapy for Acute Motor Axonal Neuropathy (AMAN) indicate notable improvements in strength, balance, and mobility [4].

Innovative treatments, while promising, can sometimes result in difficulties or negative effects. For example, because of the possibility of neurotoxicity, patients receiving large doses of methotrexate for Primary Central Nervous System Lymphoma (PCNSL) need to be closely monitored. Even while dopaminergic treatments for Parkinson's disease are successful, they can occasionally cause motor problems that need further control techniques. Risks associated with intrathecal therapy, such as those for spinal muscular atrophy, include unfavorable immunological reactions and local inflammation. In order to minimize potential consequences, these cases highlight the significance of individualized approaches, close observation, and prompt intervention [5].

The emerging trend of combination therapies aimed at synergistically targeting multiple disease pathways. Through a series of case studies and clinical vignettes, we will illustrate the practical application of these principles in real-world patient care scenarios. Despite the considerable progress made in CNS drug discovery and development, significant challenges persist, posing formidable barriers to therapeutic success. These challenges include the inherent complexity of CNS biology, the risk of off-target effects and toxicity, poor patient compliance and adherence to treatment regimens, and the stringent regulatory requirements governing CNS drug approval [6].

Neurological Disorders and Therapeutic Targets

Neurological disorders impose a significant burden on global health, accounting for 3% of the worldwide disease burden. Key neurological disorders include Alzheimer's, Parkinson's disease, multiple sclerosis, epilepsy, and headache disorders (migraine, tension-type headache, and medication-overuse headache). Despite representing a seemingly small overall percentage, conditions like dementia, epilepsy, migraine, and stroke rank among the top 50 causes of disability-adjusted life years (DALYs). Migraine and epilepsy are particularly impactful, comprising one-third and one-fourth of the neurological burden, respectively. Dementia and Parkinson's disease have shown a substantial increase in burden over the past decade [7].

Neurological disorders, such as epilepsy, dementia, and headache disorders, present significant global health challenges, accounting for 3% of the worldwide disease burden. Epilepsy, characterized by recurrent unprovoked seizures, has higher prevalence and incidence rates in low- and middle-income countries (LMICs) due to factors like endemic conditions (*e.g.*, malaria), injuries, and inadequate healthcare infrastructure [8]. The median incidence in LMICs is 81.7 per 100,000 annually compared to 45.0 in high-income countries (HICs). Premature mortality from epilepsy is the highest in LMICs, driven by preventable causes such as status epilepticus and accidents. Globally, epilepsy is the 36th leading cause of disability-adjusted life years (DALYs) and the 20th leading cause of years lived with disability (YLDs), second to migraine among brain disorders [9].

Dementia, affecting cognitive functions and daily activities, poses a growing burden, particularly in aging populations. Approximately 47 million people had dementia in 2015, with projections indicating this will nearly triple by 2050, predominantly in LMICs. Despite its increasing prevalence, dementia is often stigmatized and misunderstood, leading to delayed diagnoses and inadequate care [10]. This gap is exacerbated by the scarcity of pharmacological and psychosocial interventions in LMICs. Dementia includes Alzheimer's disease, vascular dementia, and other forms, with Alzheimer's accounting for 50-60% of late-life cases. The incidence of dementia doubles every five years after age 65, highlighting the impact of aging populations on the global burden [11].

Headache disorders, including migraine, tension-type headaches (TTH), and medication-overuse headaches (MOH), are major contributors to global disability. Migraine is characterized by episodic attacks involving headache, nausea, and photophobia, often hormonally influenced in women. Chronic migraine, occurring on 15 or more days per month, is particularly debilitating. TTH, common in teenage years, lacks migraine's specific features but is similarly disruptive [12]. MOH arises from the excessive use of headache medications, making it both avoidable and highly disabling. The global one-year prevalence of migraine is 14.7%, and TTH is 20.8% among adults, with chronic headaches affecting 3% of the population. Although migraine prevalence is generally lower in Asia, studies, such as in Zambia, show high rates of headache disorders with significant economic impact [13].

These neurological conditions highlight the need for improved healthcare infrastructure, especially in LMICs, where the burden is rising rapidly due to demographic changes and insufficient resources. Effective interventions and an increased understanding of these disorders' epidemiology and risk factors are essential to mitigate their global impact [14].

Importance of Bioactive Compounds in CNS Disorders

Research into the therapeutic potential of phenolic and flavonoid compounds, carotenoids, essential oils, fatty acids, and phytosterols in neurodegenerative diseases has unveiled promising avenues for intervention. Phenolic compounds, especially flavonoids, are renowned for their robust antioxidant properties, with studies indicating their ability to mitigate the risk of cardiovascular diseases, cancers, and neurodegenerative disorders such as Parkinson's Disease (PD), Alzheimer's Disease (AD), and Motor Neuron Disease (MND). Compounds like quercetin, naringenin, and curcumin have shown neuroprotective effects in PD by preserving dopaminergic cells, while extracts from plants like Ginkgo biloba and ginseng root exhibit promise in reducing motor deficits in MND [15]. In AD, phenolic extracts from plants like *Sesamia cretica* and Adiantum *capillus-veneris* have demonstrated inhibitory effects on enzymes associated with disease progression. Carotenoids such as lycopene, zeaxanthin, and lutein have demonstrated neuroprotective effects against PD and AD, with lycopene showing the potential to reduce oxidative stress and improve cognitive function in AD models. Essential oils derived from plants like *Aloysia citrodora* and *Pinus halepensis* exhibit anti-inflammatory and antioxidant properties, showcasing neuroprotective effects against oxidative stress and amyloid-induced neurotoxicity in AD models [16]. Omega-3 Polyunsaturated Fatty Acids (PUFAs) demonstrate therapeutic potential across PD, MND, and AD, exerting anti-inflammatory and neuroprotective effects, while medium-chain fatty acids and hydroxybutyrate show promise in improving locomotor abilities in MND models. Phytosterols, resembling cholesterol in structure, show potential for modulating cholesterol metabolism in the brain and reducing amyloid production in AD models, thus holding promise in preventing hypercholesterolemia and age-related cognitive decline. These findings collectively underscore the multifaceted therapeutic potential of natural compounds in combating neurodegenerative diseases, providing a rich landscape for further exploration and therapeutic development [17].

Biomarkers in Neurodegenerative Disorders

Biomarkers in neurodegenerative disorders are measurable indicators that can provide insights into disease presence, progression, and therapeutic response. They are crucial in early diagnosis, tracking disease progression, and evaluating the effectiveness of treatments. These biomarkers can be derived from various sources, such as Cerebrospinal Fluid (CSF), blood, imaging techniques, or genetic testing. A meta-analysis examines circulating microRNAs (miRNAs) as potential non-invasive biomarkers for diagnosing Parkinson's Disease (PD) through a meta-analysis of 44 studies involving 3298 PD patients and 2529 controls. The pooled results indicate that circulating miRNAs achieve good diagnostic accuracy, with a sensitivity of 79%, specificity of 82%, and an area under the curve (AUC) of 0.87. Notably, miRNA clusters demonstrated higher diagnostic performance compared to single miRNAs. Subgroup analyses revealed better accuracy in studies from Asian populations and when using serum samples. While miRNAs offer promise due to their stability and ability to cross the blood-brain barrier, inconsistencies in diagnostic values due to heterogeneity and study designs necessitate further validation [18].

Another study evaluated the INNOTEST® PHOSPHO-TAU(181P) assay for its ability to differentiate Alzheimer's Disease (AD) from dementia with Lewy bodies (DLB) and age-matched controls (CS). CSF levels of tau phosphorylated at threonine 181 (P-tau181P) were significantly higher in AD patients compared to DLB and CS groups. Among the biomarkers analyzed (P-tau181P, T-tau, and Aβ1-42), P-tau181P was the most statistically significant for distinguishing AD from DLB, as confirmed through discriminant analysis, classification tree, and logistic regression. The assay demonstrated strong analytical performance, including selectivity, precision, and robustness, highlighting its potential utility in clinical differential diagnosis of AD and DLB [19].

Research assessed three CSF biomarkers—sAPPβ, YKL-40, and NfL—across Amyotrophic Lateral Sclerosis (ALS), frontotemporal dementia (FTD), and controls. YKL-40 levels increased, and sAPPβ levels decreased in ALS and FTD compared to controls, with the lowest sAPPβ and sAPPβ/YKL-40 ratios observed in FTD. In FTD, these markers correlated with disease severity. Across the ALS-FTD spectrum, NfL and NfL:sAPPβ ratios correlated with cognitive performance. In ALS, YKL-40 levels were linked to faster progression and shorter survival. The sAPPβ/YKL-40 ratio positively correlated with cortical thickness in frontotemporal regions, highlighting these biomarkers as valuable tools for staging and prognosis in ALS-FTD [20].

One more study evaluated Cerebrospinal Fluid (CSF), total α-synuclein (t-α-syn), and total tau (t-tau) as biomarkers for Parkinson's Disease (PD), dementia with Lewy bodies (DLB), and Alzheimer's Disease (AD) in newly diagnosed patients. Using validated methods, lower t-α-syn levels were observed in PD, DLB, and AD patients compared to controls (16%, 23%, and 26% reductions, respectively). However, t-α-syn levels did not significantly differ between the three disorders. The t-tau/t-α-syn ratio demonstrated improved diagnostic performance compared to single markers. These findings suggest that t-α-syn, alongside tau, could enhance CSF biomarker panels for differential diagnosis of PD, DLB, and AD [21].

Blood-based biomarkers emphasize PD's complexity and heterogeneity, highlighting challenges in diagnosis, disease progression assessment, and treatment evaluation. Biomarkers like alpha-synuclein, amyloid-beta, tau proteins, neurofilament light chain (NfL), small non-coding RNAs, and inflammatory markers are evaluated for their diagnostic and prognostic utility. Despite promising research, no single biomarker has been universally accepted or clinically implemented. The review suggests combining multiple biomarkers with advanced computational methods could provide a multidimensional approach for better diagnosis, disease tracking, and personalized therapies. The article also underscores the need for standardized methodologies and large-scale studies to enhance biomarkers' reliability and applicability [22].

Combination Therapy for Neurodegenerative Disorders

Combination therapy for neurodegenerative disorders targets multiple disease pathways simultaneously to enhance efficacy. For Alzheimer's, options include cholinesterase inhibitors with NMDA antagonists or anti-inflammatory agents with antioxidants. Parkinson's treatments combine dopaminergic drugs with MAO-B inhibitors or deep brain stimulation with levodopa. ALS therapies like riluzole with edaravone address oxidative stress, while Huntington's combines antisense oligonucleotides with symptomatic treatments. Challenges include drug interactions, dosage optimization, and patient compliance [23].

A randomized, double-blind trial evaluated the safety and efficacy of Cerebrolysin, donepezil, and their combination in mild-to-moderate Alzheimer's Disease (AD) patients (MMSE score 12–25). Over 28 weeks, global outcomes (CIBIC+), cognition (ADAS-cog+), functioning (ADCS-ADL), and behavior (NPI) were assessed. While cognitive, functional, and behavioral improvements showed no significant differences across groups, global outcomes favored Cerebrolysin and combination therapy. Cognitive performance improved in all groups, with the combination therapy showing the greatest benefits. The findings suggest that Cerebrolysin is as effective as donepezil and that combining neurotrophic and cholinergic treatments is safe, warranting further exploration of long-term synergistic effects [24].

A randomized, double-blind, placebo-controlled, parallel-group study evaluated the efficacy and tolerability of atomoxetine (ATX) in improving cognitive performance in patients with mild-to-moderate Alzheimer's disease. The study involved 92 participants aged 55 and older, who received ATX (25-80 mg/day) or placebo (PLA) alongside ongoing cholinesterase-inhibitor therapy for six months. While the primary measure, Alzheimer's Disease Assessment Scale—Cognitive Portion (ADAS-Cog), did not show significant differences between groups, secondary safety measures revealed increased heart rate, diastolic blood pressure, and weight decrease in the ATX group compared to PLA. ATX was generally well tolerated but did not significantly enhance cognitive function [25].

An invention focuses on combination therapy for treating neurodegenerative diseases using isoliquiritigenin in combination with flupirtine or mirtazapine. The combination aims to prevent or treat neuronal apoptosis, a key feature of degenerative brain conditions such as Alzheimer's disease, Parkinson's disease, and stroke. By addressing neuronal cell death and spinal cord impairment, the combination of isoliquiritigenin with flupirtine or mirtazapine enhances therapeutic efficacy. The compositions and methods are designed for neurological diseases, including dementia, Huntington's disease, multiple sclerosis, and acute brain injuries, and can be administered orally with pharmaceutically acceptable carriers for improved treatment outcomes [26].

Clinical Strategies and Management Approaches

Invasive approaches to CNS drug delivery offer the advantage of directly administering a broad range of formulations, including small and large molecules, either alone or in combination, *via* Intracerebral (IC) or Intracerebroventricular (ICV) routes. Techniques such as intracerebral implants, ICV infusion, Convection-Enhanced Delivery (CED), and polymer or microchip systems allow precise delivery of therapeutics to the brain, bypassing the Blood-Brain Barrier

(BBB) [27]. However, these methods come with significant drawbacks, including high hospitalization costs, the need for anesthesia, potential disruption of the BBB leading to the spread of cancerous cells entry of unwanted blood components, and the risk of lasting neuronal damage. Specific techniques like osmotic disruption and MRI-guided focused ultrasound temporarily open the BBB to enhance drug delivery, but they also pose risks such as unwanted drug delivery to healthy brain tissue, increased intracranial pressure, and physiological stress. Despite their potential, these invasive methods require careful consideration of their risks and benefits. Non-invasive approaches to CNS drug delivery, including adsorption-mediated drug transport mechanism (AMT), carrier-mediated drug transport mechanism (CMT), and receptor-mediated drug transport mechanism (RMT), offer promising strategies for targeted therapy. Carrier-mediated transport mechanisms facilitate the entry of essential nutrients into the brain, while receptor-mediated delivery systems, such as Transferrin Receptor (TR) and Insulin Receptor (IR) mediated transcytosis, enable the targeted delivery of drugs across the Blood-Brain Barrier (BBB) [28]. Biological approaches utilize antibodies and ligands to target specific receptors, while chemical modifications, such as prodrug strategies, enhance drug penetration and stability. Colloidal approaches employ liposomes, niosomes, nanoparticles, Solid Lipid Nanoparticles (SLNs), nanoemulsions, and microemulsions to improve drug distribution, efficacy, and safety. These diverse non-invasive techniques hold promise for overcoming the challenges of CNS drug delivery and improving therapeutic outcomes. Iontophoretic delivery utilizes external electric fields to transport ionized molecules across the Blood-Brain Barrier (BBB), offering a promising method for targeted CNS drug delivery. This technique has shown potential for delivering proteins and peptides into the brain, with programmable devices allowing controlled manipulation of drug delivery [29]. The focused Ultrasound (FUS) Technique employs ultrasound in combination with gas microbubbles to temporarily enhance BBB permeability, facilitating drug delivery into the brain. FUS concentrates sound energy on a small target volume in the brain with minimal damage to surrounding tissue, making it a promising approach for non-invasive drug delivery. Intranasal delivery, also known as olfactory delivery, offers a route for direct drug targeting to the brain *via* the olfactory pathway. Drug absorption through the nasal respiratory epithelium can occur through various mechanisms, including paracellular transport, transcellular absorption, and carrier-mediated transport, providing an effective means of delivering drugs to the CNS [30].

Effective management of CNS disorders requires a multifaceted approach that includes rational drug selection, personalized medicine, combination therapies, careful monitoring, and strategies to enhance patient adherence. Rational drug selection involves choosing drugs based on their ability to target specific disease mechanisms, considering pharmacokinetics, patient-specific factors, and potential interactions to maximize efficacy and minimize side effects [31]. Personalized medicine tailors treatments to individual genetic, biomarker, and phenotypic profiles, improving outcomes by accounting for the heterogeneity of CNS disorders. For example, genetic testing and biomarkers guide therapy in conditions like Alzheimer's and Parkinson's diseases, while pharmacogenomics informs dosage adjustments to enhance drug efficacy and reduce side effects. Combination therapies, which use multiple drugs or modalities, aim to achieve synergistic effects, addressing the complex nature of CNS diseases. This approach can improve outcomes in disorders like Parkinson's, multiple sclerosis, and schizophrenia by targeting different pathways and reducing individual drug doses and side effects [32]. Monitoring off-target effects and toxicity is crucial due to the brain's sensitive nature, involving regular clinical assessments, therapeutic drug monitoring, and advanced imaging techniques. Educating patients on side effects and encouraging timely reporting, along with dosage adjustments or alternative therapies, helps manage adverse effects. Enhancing patient adherence is essential, particularly given the chronic nature of CNS disorders and complex medication regimens. Strategies to improve adherence include simplifying regimens, using long-acting formulations, and incorporating digital health technologies like reminders and telemedicine [33]. Patient education on adherence importance, managing expectations, and addressing barriers like side effects and financial constraints are crucial. Engaging patients through shared decision-making fosters responsibility and improves adherence, supported by healthcare providers, family, and peer groups. These comprehensive strategies leverage pharmacological advancements, personalized medicine, and patient-centered care to effectively manage CNS disorders and enhance therapeutic outcomes [34].

Case Studies on Neurological Disorders

The following Table **1** includes case studies related to various neurological disorders, such as Sclerosis, Myasthenia gravis, Parkinson's, *etc.*

Table 1. Some findings related to neurological disorders.

Sr. No.	Disease Target	Patient Detail	Test Result	Drug Used	Result of Case Study	Refs.
01.	Amyotrophic Lateral Sclerosis (ALS)	24-year-old male with ALS. DOB: 1996/06/02 and the Patient had tonsillectomy at age 10, No known family history of ALS. The patient's father has asthma; Mother has Type II diabetes.	Upper Motor Neurons (UMN) -Positive Babinski's sign bilaterally . Positive Clonus Test bilaterally, Lower Motor Neurons (LMN) -Increased Achilles reflex - Grade 4+, Increased patellar reflex - Grade 3+ Increased biceps reflex - Grade 3+ Increased triceps reflex - Grade 3+	Riluzole, 100mg orally/day, Acetaminophen as needed.	No significant changes in most Jared's results/measures compared to six weeks ago, having stable/consistent scores in many of the outcome measures were due to the progressive nature of ALS.	[35]
02.	Acute Motor Axonal Neuropathy (AMAN) is a variant type of Guillain-Barre Syndrome (GBS)	Trish Jones, 62 year old female	FUNCTION AND MOBILITY TESTING	High dose of intravenous immunoglobulin (ivig) therapy as needed.	After 12 weeks of physiotherapy sessions, Trish has improved significantly on multiple components of her diagnosis that have impacted her independence the most, such as her strength, range of motion, balance, and gait.	[36]
03.	Primary Central Nervous System Lymphoma (PCNSL)	18-year-old male	Tumor cells positive for leukocyte common antigen, anaplastic lymphoma kinase (ALK), CD30, and CD4, *etc.*	High-dose methotrexate with leucovorin rescue, intrathecal methotrexate, vincristine, procarbazine, and dexamethasone.	The patient will be on prophylactic antiepileptic medication and had no further seizures after the initial surgery.	[37]
04.	Myasthenia gravis	35 year old woman (librarian)	A blood test found high levels of acetylcholine receptor antibodies and diagnosed her with Myasthenia Gravis (MG), class iia, *etc.*	Cholinesterase inhibitor	6-week up/Discharge, Patient's score improved QoL to 157/30 effectively.	[38]

(Table 1) cont.....

Sr. No.	Disease Target	Patient Detail	Test Result	Drug Used	Result of Case Study	Refs.
05.	Parkinson's Disease (PD), progressive degeneration dopaminergic nerves	Mrs. Martina Tribianni is a 66-year-old female, elementary school teacher and ballet teacher.	Diagnosed with stage 3 idiopathic PD, *etc.*	Dopaminergic therapy (Levodopa/Car bidopa)	Mrs. Tribbiani seemed to really benefit from the exercise programs that were designed for her and will continue to participate in them to further improve her strength, gait, balance, and ROM.	[39]
06.	Post-polio syndrome (PPS)	40 year old male, patient contracted poliomyelitis when he was 15 years old.	Negative special tests for bilateral shoulder impingement: Neers, Hawkins-Kenn edy, Supraspinatus "Full/Empty Can" Test Negative Drop Arm test bilaterally, Positive "Hornblowers" and "Lift Off" test, *etc.*	Not reported	The patient was able to make significant impacts on key indicators including fatigue, levels of pain and disability, and improvements in upper and lower extremity strength.	[40]
07.	Myotonic dystrophy type 1 (DM1)	23 individuals with juvenile or adult phenotypes of DM1(2016-2019).	Mini-bestest, the 10-m walk test, the 6-min walk test, *etc.*	Not reported	Men lost more MIMS than women, and the adult phenotype lost more MIMS than the juvenile phenotype.	[41]
08.	Guillain-Barre Syndrome (Sub-acute)	Tom Brown, 56 year old male	Negative Manual Muscle Testing (MMT), Gait assessment(unable to complete walk), Pulmonary function tests(restriction in expiration capability and muscle weakness), *etc.*	Not reported	Tom's symptoms from GBS have significantly decreased since starting physical therapy.	[42]

(Table 1) cont.....

Sr. No.	Disease Target	Patient Detail	Test Result	Drug Used	Result of Case Study	Refs.
09.	Myotonic Dystrophy Type 1 (DM1)	30-year-old Mr. J	Manual Muscle Tests-3-/5 for neck extension and 2-/5 for ankle dorsiflexion bilaterally, Berg balance test - 43/56 as a total score, *etc.*	Not reported	At the end of his 6-week intervention, Mr. J was feeling more confident in his community ambulation and slightly stronger. He was grateful for the knowledge he learned about his condition and the things that could be done to slow its progression.	[43]
10.	Charcot-Marie-Tooth Disease (CMT)	18 year old male	Myotomes: decrease strength through L4- S1 bilaterally, Dural testing: negative Spurling's Test, Straight Leg Raise Test, and Slump Test, Reflexes: negative Babinski and Hoffman's, decreased deep tendon reflexes bilaterally (Achilles and patellar), *etc.*	Not reported	Every 3 months, the progression of the disease is assessed, and the outcome measures are agan tested to prescribe new exercises and self-management to ensure maximal retention of function. No issues were found with the upper extremities but would be monitored during follow-up sessions using dynamometry, 9-hole peg test, CMT neuropathy score, and ONLS to assess changes in function.	[44]

(Table 1) cont.....

Sr. No.	Disease Target	Patient Detail	Test Result	Drug Used	Result of Case Study	Refs.
11.	Trigeminal Neuralgia (TN)	Ms. R, 35-year-old female	Magnetic Resonance Imaging (MRI) demonstrates unilateral neurovascular compression with mild morphologic changes of the nerve root; contact observed at root entry, *etc.*	Escitalopram Oxalate (10 mg/day); Gabapentin (300 mg/day)	The physiotherapy treatment used in this case focuses on electrophysical agents, soft tissue massage, and education about self-management techniques which provided some relief. Integrating physiotherapy treatments into a multidisciplinary approach can assist in providing optimal outcomes for patients with TN.	[45]
12.	Right-side ischemic cerebellar stroke	63 year old male (smoker)	Positive limb matching test, gait observation, the BERG score of 44/56, a TUG time of 22.2 seconds, and an ABC score of 77/100, *etc.*	Not reported	The patient has a good prognosis to achieve functional long-term goals if he continues to complete his rehabilitation therapy.	[46]
13.	Sub-cortical vascular dementia (Binswanger's Disease)	Age: 60 y/o Gender: Female, Occupation: Retired secretary for the government.	Frontal Assessment Battery (FAB) test to differentiate between Alzeihmer's disease and Vascular dementia, CT scan of the brain(thickening and narrowing (atherosclerosis) of arteries), *etc.*	Prinivil 10 mg once a day for high blood pressure, Daily women's multivitamin (One a Day Women's 50+)	Care will be provided indefinitely.	[47]

(Table 1) cont.....

Sr. No.	Disease Target	Patient Detail	Test Result	Drug Used	Result of Case Study	Refs.
14.	Cerebral Palsy (CP)	24-year-old male (diagnosed at the age of 5 at Level 1 of the GMFCS score)	Medical history of depression (3 years) as well as a history of cancer in his immediate family, Standing Heel Raise: L 20, R 10, FABER (-) FADDIR (-) Hip scouring (-), Second appointment: Thomas (-) Thompson (-), *etc.*	Paroxetine (25mg orally once a day)	Improvement in right side ankle dorsiflexion, a decrease in foot drop, improved glute strength, a decrease in hip pain at rest and during activity, as well as a decrease in his Fatigue Severity Scale score.	[48]
15.	Benign paroxysmal positional vertigo (BPPV)	The patient is a 55-year-old male(profession-accountant)	Baseline scores of 18/24 on the Dynamic Gait Index, 37/100 on the Dizziness Handicap Inventory, 76% on the Vertigo Handicap Questionnaire, and 44% on the Activities-specific Balance Confidence Scale, *etc.*	40mg of Diovan daily(to control B.P.), 5mg of Crestor daily.	The physiotherapist will be the main health care professional responsible in this case, otolaryngologists may also be used if imaging or surgery is required.	[49]
16.	Multiple System Atrophy-Parkinsonian and Cerebellar (MSA-P and MSA-C)	Mr. L is a 62-year-old retired NASA astronaut	MRI: T2 weighted MRI revealed a hyperintense putaminal slit sign bilaterally, *etc.*	Levodopa (250 mg 4X/day, for 4 months) ,Levodopa (500 mg 4X/day, for 2 months) CBD for pain management (occasional)	There are still some concerns for Mr. L's safety as he still suffers from slight gait ataxia and balance impairments due to the nature of the disease.	[50]
17.	Neuralgic amyotrophy (NA) - Parsonage-Turner Syndrome(Brachial neuritis)	13-year-old right-handed African-American (reported injuring his right shoulder while doing a push-up in gym class)	Electromyography and nerve conduction studies indicated neuralgic amyotrophy, *etc.*	Not reported	PF-10=100 Quick DASH=2.27% Disability	[51]

(Table 1) cont.....

Sr. No.	Disease Target	Patient Detail	Test Result	Drug Used	Result of Case Study	Refs.
18.	Traumatic brain injury (TBI)	The client is a 33-year-old female(report of physical violence by her partner).	MRI showed intracranial contusion affecting the right parietal and temporal lobes. Multiple micro traumas were also visible in the image, which was interpreted as multiple mtbis, likely caused by previous IPV, *etc.*	Sertraline (Zoloft), Aspirin (for frequent headaches)	She was discharged to her mother's home, the patient will be completing home physiotherapy, working with a social worker and psychologist.	[52]
19.	Progressive supranuclear palsy (PSP)	Jim is a 67 year old retired farmer that lives with his wife Julia	Babinski's sign (+),Clonus (-) Hoffman's (-)	Hydrochlorothiazide, Atorvastatin (Lipidor), Acebutolol (Sectral)	After 16 weeks, re-assessment revealed improvements in the patient's Progressive Supranuclear Palsy Rating Scale (PSP-RS) score, Timed Up and Go (TUG) score, and Berg Balance Scale when compared to baseline.	[53]
20.	Locked-In Syndrome (LIS)	Mr. Gary is a 38-year-old male, landlord of multiple apartment buildings.	FIM is an 18-item test, RAND 36-Item Health Survey	Not reported	The physiotherapy team is working to improve his strength and ROM through RST, beginning with 60 minute sessions 5-6 times per week.	[54]

Challenges in Targeting the CNS

Drug distribution in the CNS faces significant challenges, primarily due to the Blood-Brain Barrier (BBB), which restricts the entry of therapeutic drugs while allowing essential nutrients and clearing toxic compounds. The BBB, formed by specialized cells and tight junctions, works alongside the meningeal/arachnoid and choroid plexus barriers, posing substantial obstacles to CNS drug delivery [55]. According to a 2018 publication, 95% of drug molecules struggle to cross these barriers, impeding CNS therapeutic development. Several novel approaches aim to overcome the BBB. Hyperosmotic solutions can disrupt the BBB, facilitating drug delivery, particularly in oncology for chemotherapy. Viral vectors, such as adeno-associated viruses, offer gene delivery capabilities but face challenges like immunogenicity and high production costs. Nanoparticles provide promising drug delivery systems due to their ability to carry payloads and control drug release, with advanced systems like nanogels showing significant advantages [56]. Alternative strategies bypass the BBB altogether. Convection-Enhanced Delivery (CED) directly infuses drugs into CNS tissue, avoiding systemic toxicity and enabling high-dose delivery to targeted areas. However, CED faces issues like backflow and white matter edema. Another emerging approach involves utilizing Cerebrospinal Fluid (CSF) to distribute drugs across the CNS. While still in preclinical stages, this method leverages the CSF's natural role in nutrient and molecule distribution, aiming to develop translatable human models for effective CNS drug delivery [57]. One major obstacle to the broad adoption of new delivery systems is the financial cost associated with their development and implementation. Advanced technologies including gene treatments, targeted ultrasound systems, and carriers based on nanotechnology have very high research and development (R&D) expenses, frequently surpassing hundreds of millions of dollars per product. These costs result from lengthy clinical trials, comprehensive preclinical research, and the strict requirements of regulatory compliance [58]. The production of adeno-associated viral vectors, which are necessary for gene therapy, for example, entails intricate and resource-intensive procedures that increase the financial burden. As a result, only a small population can afford these treatments due to their unreasonably high cost per unit. Costs are further increased by the infrastructure needed for these distribution systems. It is expensive to set up and maintain state-of-the-art facilities with sterile settings for the synthesis of nanoparticles or sophisticated equipment for methods like convection-enhanced delivery (CED) [59]. Low- and middle-income nations (LMICs), whose healthcare resources are constrained, are disproportionately impacted by these expenses. The financial strain on insurance companies and healthcare systems is significant, even in high-income nations. The economic feasibility of these treatments is further complicated by concerns regarding their long-term cost-effectiveness, including their durability, the need for retreatment, and the handling of adverse effects (Fig. **1**).

Future Perspectives in CNS Drug Discovery and Development

Ischemic stroke therapy initially relied on rodent models treated at the time of infarction, later shifting to 1-2 hours post-occlusion in young rats, mirroring clinical trial designs. However, these models failed to predict human outcomes, partly due to discrepancies in age, species, and stroke characteristics, and neglect of comorbidities. Recognizing these limitations, researchers now advocate for more representative models, including aged and gyrencephalic animals, despite higher costs and time requirements [60]. Improving drug discovery involves embracing alternative hypotheses and learning from clinical failures. Better preclinical models with higher face validity are essential. Effective predictions require extensive and time-consuming testing in models that closely mimic human disease. Understanding disease mechanisms deeply and avoiding oversimplification in target modulation is critical. Ion channels, for instance, exhibit complex roles beyond ion flux, influencing processes like protein phosphorylation and cell adhesion. Therefore, drug discovery must account for these multifaceted functions. Combination therapies offer promise, as seen in Cystic Fibrosis (CF) treatment. CF therapy evolved from unsuccessful gene therapy to the effective use of both an activator and a chaperone, targeting different aspects of the defective CFTR protein. Similarly, CNS diseases, given their complexity, may benefit from multi-target approaches [61]. The FDA has been cautious but is becoming more open to combination therapies for neurodegenerative diseases, provided there is strong mechanistic justification and safety. Longer clinical trials may be necessary, especially for diseases like Alzheimer's (AD), where early intervention could prevent irreversible damage. Early-stage treatments, identified through advanced biomarkers, might require prolonged administration in otherwise healthy individuals. This necessitates economically viable models for extensive longitudinal studies [62].

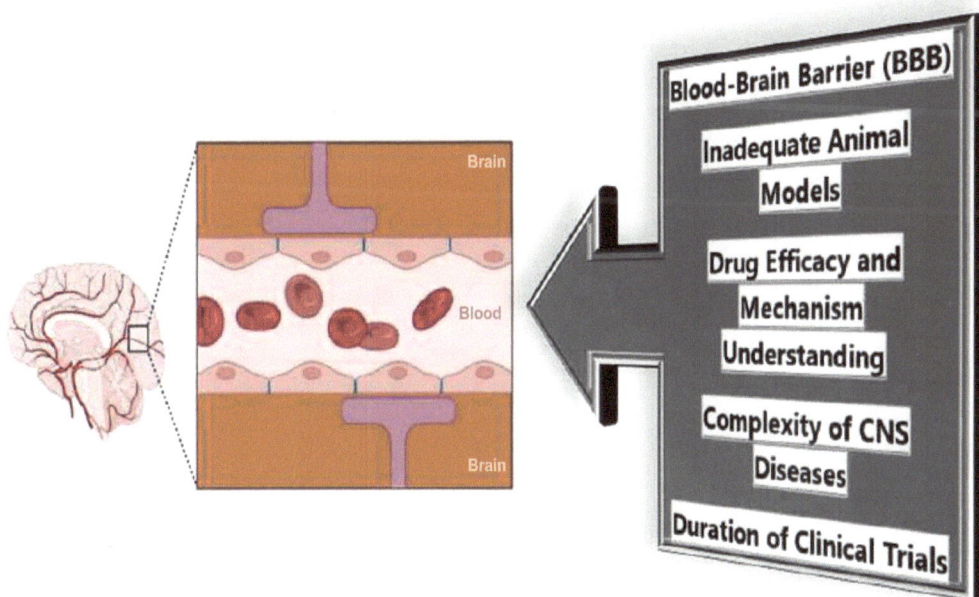

Fig. (1). Challenges in targeting CNS.

CONCLUSION

The Central Nervous System (CNS) is critical for regulating essential physiological processes, making the treatment of its disorders imperative for maintaining overall health and well-being. Neurological disorders such as Alzheimer's, Parkinson's, and depression pose significant challenges due to their complex pathophysiology and the difficulty of delivering therapeutics across the Blood-Brain Barrier (BBB). Bioactive compounds offer promising therapeutic potential by targeting specific molecular mechanisms involved in these disorders. However, effective CNS drug delivery remains challenging due to the restrictive nature of the BBB, necessitating innovative approaches like nanoparticles, viral vectors, and alternative delivery methods such as convection-enhanced delivery and cerebrospinal fluid utilization. Future research should focus on developing better preclinical models that closely mimic human disease and exploring multi-target approaches. Longer clinical trials and early intervention strategies, supported by advanced biomarkers, are crucial for preventing irreversible damage in diseases like Alzheimer's. Overall, a comprehensive and innovative approach to CNS drug discovery and delivery is essential for addressing the global burden of neurological disorders.

REFERENCES

[1]   Ludwig, P.E.; Reddy, V.; Varacallo, M. *Neuroanatomy, Central Nervous System*; CNS, **2023**.

[2]   Palanisamy, C.P.; Pei, J.; Alugoju, P.; Anthikapalli, N.V.A.; Jayaraman, S.; Veeraraghavan, V.P.; Gopathy, S.; Roy, J.R.; Janaki, C.S.; Thalamati, D.; Mironescu, M.; Luo, Q.; Miao, Y.; Chai, Y.; Long, Q. New strategies of neurodegenerative disease treatment with extracellular vesicles (EVs) derived from mesenchymal stem cells (MSCs). *Theranostics,* **2023**, *13*(12), 4138-4165.
[http://dx.doi.org/10.7150/thno.83066] [PMID: 37554286]

[3]   Upadhyay, R.K. Drug delivery systems, CNS protection, and the blood brain barrier. *BioMed Res. Int.,* **2014**, *2014*, 1-37.
[http://dx.doi.org/10.1155/2014/869269] [PMID: 25136634]

[4]   Biebl, J.T.; Azqueta-Gavaldon, M.; Wania, C.; Zettl, O.; Woiczinski, M.; Bauer, L.; Storz, C.; Bötzel, K.; Kraft, E. Resistance Training Combined with Balance or Gait Training for Patients with Parkinson's Disease: A Randomized Controlled Pilot Study. *Parkinsons Dis.,* **2022**, *2022*, 1-7.
[http://dx.doi.org/10.1155/2022/9574516] [PMID: 36247107]

[5]   Choi, Y.S. Recent advances in the management of primary central nervous system lymphoma. *Blood Res.,* **2020**, *55*(S1), S58-S62.
[http://dx.doi.org/10.5045/br.2020.S010] [PMID: 32719178]

[6]   Gad, S.C. Safety and regulatory requirements and challenge for CNS drug development. *Neurobiol. Dis.,* **2014**, *61*, 39-46.
[http://dx.doi.org/10.1016/j.nbd.2013.09.017] [PMID: 24090873]

[7]   Perjoc, R.S.; Roza, E.; Vladacenco, O.A.; Teleanu, D.M.; Neacsu, R.; Teleanu, R.I. Functional Neurological Disorder–Old Problem New Perspective. *Int. J. Environ. Res. Public Health,* **2023**, *20*(2), 1099.
[http://dx.doi.org/10.3390/ijerph20021099] [PMID: 36673871]

[8]   Gilmour, G.S.; Nielsen, G.; Teodoro, T.; Yogarajah, M.; Coebergh, J.A.; Dilley, M.D.; Martino, D.;

Edwards, M.J. Management of functional neurological disorder. *J. Neurol.,* **2020**, *267*(7), 2164-2172.
[http://dx.doi.org/10.1007/s00415-020-09772-w] [PMID: 32193596]

[9]   Levira, F.; Thurman, D.J.; Sander, J.W.; Hauser, W.A.; Hesdorffer, D.C.; Masanja, H.; Odermatt, P.; Logroscino, G.; Newton, C.R. Premature mortality of epilepsy in low- and middle-income countries: A systematic review from the Mortality Task Force of the International League Against Epilepsy. *Epilepsia,* **2017**, *58*(1), 6-16.
[http://dx.doi.org/10.1111/epi.13603] [PMID: 27988968]

[10] Livingston, G.; Huntley, J.; Sommerlad, A.; Ames, D.; Ballard, C.; Banerjee, S.; Brayne, C.; Burns, A.; Cohen-Mansfield, J.; Cooper, C.; Costafreda, S.G.; Dias, A.; Fox, N.; Gitlin, L.N.; Howard, R.; Kales, H.C.; Kivimäki, M.; Larson, E.B.; Ogunniyi, A.; Orgeta, V.; Ritchie, K.; Rockwood, K.; Sampson, E.L.; Samus, Q.; Schneider, L.S.; Selbæk, G.; Teri, L.; Mukadam, N. Dementia prevention, intervention, and care: 2020 report of the Lancet Commission. *Lancet,* **2020**, *396*(10248), 413-446.
[http://dx.doi.org/10.1016/S0140-6736(20)30367-6] [PMID: 32738937]

[11] George-Carey, R.; Adeloye, D.; Chan, K.Y.; Paul, A.; Kolčić, I.; Campbell, H.; Rudan, I. An estimate of the prevalence of dementia in Africa: A systematic analysis. *J. Glob. Health,* **2012**, *2*(2), 020401.
[http://dx.doi.org/10.7189/jogh.02.020401] [PMID: 23289076]

[12] Ahmed, F. Headache disorders: differentiating and managing the common subtypes. *Br. J. Pain,* **2012**, *6*(3), 124-132.
[http://dx.doi.org/10.1177/2049463712459691] [PMID: 26516483]

[13] Kristoffersen, E.S.; Lundqvist, C. Medication-overuse headache: epidemiology, diagnosis and treatment. *Ther. Adv. Drug Saf.,* **2014**, *5*(2), 87-99.
[http://dx.doi.org/10.1177/2042098614522683] [PMID: 25083264]

[14] Carroll, W.M. The global burden of neurological disorders. *Lancet Neurol.,* **2019**, *18*(5), 418-419.
[http://dx.doi.org/10.1016/S1474-4422(19)30029-8] [PMID: 30879892]

[15] Sun, W.; Shahrajabian, M.H. Therapeutic Potential of Phenolic Compounds in Medicinal Plants—Natural Health Products for Human Health. *Molecules,* **2023**, *28*(4), 1845.
[http://dx.doi.org/10.3390/molecules28041845] [PMID: 36838831]

[16] Chohra, D.; Ferchichi, L.; Cakmak, Y.S.; Zengin, G.; Alsheikh, S.M. Phenolic profiles, antioxidant activities and enzyme inhibitory effects of an Algerian medicinal plant (Clematis cirrhosa L.). *S. Afr. J. Bot.,* **2020**, *132*, 164-170.
[http://dx.doi.org/10.1016/j.sajb.2020.04.026]

[17] Shahidi, F.; Ambigaipalan, P. Omega-3 Polyunsaturated Fatty Acids and Their Health Benefits. *Annu. Rev. Food Sci. Technol.,* **2018**, *9*(1), 345-381.
[http://dx.doi.org/10.1146/annurev-food-111317-095850] [PMID: 29350557]

[18] Zhang, W.T.; Wang, Y.J.; Yao, Y.F.; Zhang, G.X.; Zhang, Y.N.; Gao, S.S. Circulating microRNAs as potential biomarkers for the diagnosis of Parkinson's disease: A meta-analysis. *Neurología (English Edition),* **2024**, *39*(7), 573-583.
[http://dx.doi.org/10.1016/j.nrleng.2024.07.004] [PMID: 39232595]

[19] Vanderstichele, H.; De Vreese, K.; Blennow, K.; Andreasen, N.; Sindic, C.; Ivanoiu, A.; Hampel, H.; Bürger, K.; Parnetti, L.; Lanari, A.; Padovani, A.; DiLuca, M.; Bläser, M.; Ohrfelt Olsson, A.; Pottel, H.; Hulstaert, F.; Vanmechelen, E. Analytical performance and clinical utility of the INNOTEST® PHOSPHO-TAU(181P) assay for discrimination between Alzheimer's disease and dementia with Lewy bodies. *Clin. Chem. Lab. Med.,* **2006**, *44*(12), 1472-1480.
[http://dx.doi.org/10.1515/CCLM.2006.258] [PMID: 17163825]

[20] Illán-Gala, I.; Alcolea, D.; Montal, V.; Dols-Icardo, O.; Muñoz, L.; de Luna, N.; Turón-Sans, J.; Cortés-Vicente, E.; Sánchez-Saudinós, M.B.; Subirana, A.; Sala, I.; Blesa, R.; Clarimón, J.; Fortea, J.; Rojas-García, R.; Lleó, A. CSF sAPPβ, YKL-40, and NfL along the ALS-FTD spectrum. *Neurology,* **2018**, *91*(17), e1619-e1628.
[http://dx.doi.org/10.1212/WNL.0000000000006383] [PMID: 30291183]

[21]   Førland, M.G.; Tysnes, O.B.; Aarsland, D.; Maple-Grødem, J.; Pedersen, K.F.; Alves, G.; Lange, J. The value of cerebrospinal fluid α-synuclein and the tau/α-synuclein ratio for diagnosis of neurodegenerative disorders with Lewy pathology. *Eur. J. Neurol.,* **2020,** *27*(1), 43-50.
[http://dx.doi.org/10.1111/ene.14032] [PMID: 31293044]

[22]   Tönges, L.; Buhmann, C.; Klebe, S.; Klucken, J.; Kwon, E.H.; Müller, T.; Pedrosa, D.J.; Schröter, N.; Riederer, P.; Lingor, P. Blood-based biomarker in Parkinson's disease: potential for future applications in clinical research and practice. *J. Neural Transm. (Vienna),* **2022,** *129*(9), 1201-1217.
[http://dx.doi.org/10.1007/s00702-022-02498-1] [PMID: 35428925]

[23]   Cummings, J.L.; Tong, G.; Ballard, C. Treatment Combinations for Alzheimer's Disease: Current and Future Pharmacotherapy Options. *J. Alzheimers Dis.,* **2019,** *67*(3), 779-794.
[http://dx.doi.org/10.3233/JAD-180766] [PMID: 30689575]

[24]   Alvarez, X.A.; Cacabelos, R.; Sampedro, C.; Couceiro, V.; Aleixandre, M.; Vargas, M.; Linares, C.; Granizo, E.; García-Fantini, M.; Baurecht, W.; Doppler, E.; Moessler, H. Combination treatment in Alzheimer's disease: results of a randomized, controlled trial with cerebrolysin and donepezil. *Curr. Alzheimer Res.,* **2011,** *8*(5), 583-591.
[http://dx.doi.org/10.2174/156720511796391863] [PMID: 21679156]

[25]   Mohs, R.C.; Shiovitz, T.M.; Tariot, P.N.; Porsteinsson, A.P.; Baker, K.D.; Feldman, P.D. Atomoxetine augmentation of cholinesterase inhibitor therapy in patients with Alzheimer disease: 6-month, randomized, double-blind, placebo-controlled, parallel-trial study. *Am. J. Geriatr. Psychiatry,* **2009,** *17*(9), 752-759.
[http://dx.doi.org/10.1097/JGP.0b013e3181aad585] [PMID: 19700948]

[26]   Combination therapy for treating neurodegenerative diseases using combination of isoliquiritigenin and flupirtine or mirtazapine, 2016.

[27]   Mo, F.; Pellerino, A.; Soffietti, R.; Rudà, R. Blood-brain barrier in brain tumors: Biology and clinical relevance. *Int. J. Mol. Sci.,* **2021,** *22*(23), 12654.
[http://dx.doi.org/10.3390/ijms222312654] [PMID: 34884457]

[28]   Mikitsh, J.L.; Chacko, A.M. Pathways for small molecule delivery to the central nervous system across the blood-brain barrier. *Perspect. Medicin. Chem.,* **2014,** *6*, PMC.S13384.
[http://dx.doi.org/10.4137/PMC.S13384] [PMID: 24963272]

[29]   Abdul Razzak, R.; Florence, G.J.; Gunn-Moore, F.J. Approaches to cns drug delivery with a focus on transporter-mediated transcytosis. *Int. J. Mol. Sci.,* **2019,** *20*(12), 3108.
[http://dx.doi.org/10.3390/ijms20123108] [PMID: 31242683]

[30]   Singh, A.; Jiménez-Gambín, S.; Konofagou, E.E. An all-ultrasound cranial imaging method to establish the relationship between cranial FUS incidence angle and transcranial attenuation in non-human primates in 3D. *Sci. Rep.,* **2024,** *14*(1), 1488.
[http://dx.doi.org/10.1038/s41598-024-51623-5] [PMID: 38233480]

[31]   Kvarnström, K.; Westerholm, A.; Airaksinen, M.; Liira, H. Factors contributing to medication adherence in patients with a chronic condition: A scoping review of qualitative research. *Pharmaceutics,* **2021,** *13*(7), 1100.
[http://dx.doi.org/10.3390/pharmaceutics13071100] [PMID: 34371791]

[32]   Singh, S.; Sarma, D.K.; Verma, V.; Nagpal, R.; Kumar, M. Unveiling the future of metabolic medicine: omics technologies driving personalized solutions for precision treatment of metabolic disorders. *Biochem. Biophys. Res. Commun.,* **2023,** *682*, 1-20.
[http://dx.doi.org/10.1016/j.bbrc.2023.09.064] [PMID: 37788525]

[33]   Kang, J.S.; Lee, M.H. Overview of therapeutic drug monitoring. *Korean J. Intern. Med. (Korean. Assoc. Intern. Med.),* **2009,** *24*(1), 1-10.
[http://dx.doi.org/10.3904/kjim.2009.24.1.1] [PMID: 19270474]

[34]   Krist, A.H.; Tong, S.T.; Aycock, R.A.; Longo, D.R. Engaging patients in decision-making and

behavior change to promote prevention. *Inf. Serv. Use,* **2017**, *37*(2), 105-122.
[http://dx.doi.org/10.3233/ISU-170826] [PMID: 28972524]

[35]   Jamrozik, Z.; Gawel, M.; Szacka, K.; Bakon, L. A case report of amyotrophic lateral sclerosis in a patient with Klippel-Feil syndrome—a familial occurrence: a potential role of TGF-β signaling pathway. *Medicine (Baltimore),* **2015**, *94*(4), e441.
[http://dx.doi.org/10.1097/MD.0000000000000441] [PMID: 25634178]

[36]   Hafer-Macko, C.; Hsieh, S.T.; Ho, T.W.; Sheikh, K.; Cornblath, D.R.; Li, C.Y.; McKhann, G.M.; Asbury, A.K.; Griffin, J.W. Acute motor axonal neuropathy: An antibody-mediated attack on axolemma. *Ann. Neurol.,* **1996**, *40*(4), 635-644.
[http://dx.doi.org/10.1002/ana.410400414] [PMID: 8871584]

[37]   Komaranchath, A.S.; Kuntegowdenahalli, L.C.; Jacob, L.A.; Amirtham, U. A rare case of primary anaplastic large cell lymphoma of the central nervous system. *J. Cancer Res. Ther.,* **2015**, *11*(4), 943-945.
[http://dx.doi.org/10.4103/0973-1482.162115] [PMID: 26881551]

[38]   Burns, T.M.; Grouse, C.K.; Wolfe, G.I.; Conaway, M.R.; Sanders, D.B. The MG-QOL15 for following the health-related quality of life of patients with myasthenia gravis. *Muscle Nerve,* **2011**, *43*(1), 14-18.
[http://dx.doi.org/10.1002/mus.21883] [PMID: 21082698]

[39]   Parkinson's Disease (PD): A Case Study. Physiopedia n.d. https://www.physio-pedia.com/Parkinson's_Disease_(PD):_A_Case_Study

[40]   Halstead, L.S. Post-Polio Syndrome. *Sci. Am.,* **1998**, *278*(4), 42-47.
[http://dx.doi.org/10.1038/scientificamerican0498-42] [PMID: 9532770]

[41]   Roussel, M.P.; Fiset, M.M.; Gauthier, L.; Lavoie, C.; McNicoll, É.; Pouliot, L.; Gagnon, C.; Duchesne, E. Assessment of muscular strength and functional capacity in the juvenile and adult myotonic dystrophy type 1 population: a 3-year follow-up study. *J. Neurol.,* **2021**, *268*(11), 4221-4237.
[http://dx.doi.org/10.1007/s00415-021-10533-6] [PMID: 33907889]

[42]   Case Study: Guillain-Barré Syndrome (Sub-Acute). Physiopedia n.d. https://www.physio-pedia.com/Case_Study:_Guillain-Barre_Syndrome_(Sub-Acute)

[43]   Case Study: Myotonic Dystrophy Type 1. Physiopedia n.d. https://www.physio-pedia.com/Case_Study_-_Myotonic_Dystrophy_Type_1

[44]   Charcot-Marie-Tooth Disease: A Case Study. Physiopedia n.d. https://www.physio-pedia.com/Charcot-Marie-Tooth_Disease:_A_Case_Study

[45]   Trigeminal Neuralgia: A Case Study. Physiopedia n.d. https://www.physio-pedia.com/Trigeminal_neuralgia:_A_case_study

[46]   The Outpatient Physical Therapy Experience of a 63-Year-Old Male with Right Cerebellar Ischemic Stroke: A Case Study. Physiopedia n.d. https://www.physio-pedia.com/The_outpatient_physical_therapy_experience_of_a_63_year_old_male_with_right_cerebellar_ischemic_stroke:_A_case_study

[47]   Subcortical Vascular Dementia: Case Study. Physiopedia n.d. https://www.physio-pedia.com/Subcortical_Vascular_Dementia:_Case_Study

[48]   Cerebral Palsy - Young Adult Case Study. Physiopedia n.d. https://www.physio-pedia.com/Cerebral_Palsy_-_Young_Adult_Case_Study

[49]   Benign Paroxysmal Positional Vertigo: A Case Study. Physiopedia n.d. https://www.physio-pedia.com/Benign_Paroxysmal_Positional_Vertigo:_A_Case_Study

[50]   Multiple System Atrophy: A Case Study. Physiopedia n.d. https://www.physio-pedia.com/Multiple_System_Atrophy:_A_Case_Study

[51]   Parsonage-Turner Syndrome: Case Study. Physiopedia n.d. https://www.physio-pedia.com/Parsonag--Turner_Syndrome:_Case_Study

[52]   Case Study: Traumatic Brain Injury and Intimate Partner Violence (IPV). Physiopedia n.d. https://www.physio-pedia.com/Case_Study:_Traumatic_Brain_Injury_and_Intimate_Partner_Violence_(IPV)

[53]   Progressive Supranuclear Palsy: A Case Study. Physiopedia n.d. https://www.physio-pedia.com/Progressive_Supranuclear_Palsy:_A_Case_Study

[54]   Qutubuddin, A.A.; Pegg, P.O.; Cifu, D.X.; Brown, R.; McNamee, S.; Carne, W. Validating the Berg Balance Scale for patients with Parkinson's disease: A key to rehabilitation evaluation. *Arch. Phys. Med. Rehabil.,* **2005**, *86*(4), 789-792.
[http://dx.doi.org/10.1016/j.apmr.2004.11.005] [PMID: 15827933]

[55]   Kalaydina, R.V.; Bajwa, K.; Qorri, B.; DeCarlo, A.; Szewczuk, M.R. Recent advances in "smart" delivery systems for extended drug release in cancer therapy. *Int. J. Nanomedicine,* **2018**, *13*, 4727-4745.
[http://dx.doi.org/10.2147/IJN.S168053] [PMID: 30154657]

[56]   Dong, X. Current strategies for brain drug delivery. *Theranostics,* **2018**, *8*(6), 1481-1493.
[http://dx.doi.org/10.7150/thno.21254] [PMID: 29556336]

[57]   D'Amico, R.S.; Aghi, M.K.; Vogelbaum, M.A.; Bruce, J.N. Convection-enhanced drug delivery for glioblastoma: a review. *J. Neurooncol.,* **2021**, *151*(3), 415-427.
[http://dx.doi.org/10.1007/s11060-020-03408-9] [PMID: 33611708]

[58]   Malik, S.; Muhammad, K.; Waheed, Y. Emerging Applications of Nanotechnology in Healthcare and Medicine. *Molecules,* **2023**, *28*(18), 6624.
[http://dx.doi.org/10.3390/molecules28186624] [PMID: 37764400]

[59]   Seo, Y.E.; Bu, T.; Saltzman, W.M. Nanomaterials for convection-enhanced delivery of agents to treat brain tumors. *Curr. Opin. Biomed. Eng.,* **2017**, *4*, 1-12.
[http://dx.doi.org/10.1016/j.cobme.2017.09.002] [PMID: 29333521]

[60]   West, F.D.; Kaiser, E.E. Large animal ischemic stroke models: replicating human stroke pathophysiology. *Neural Regen. Res.,* **2020**, *15*(8), 1377-1387.
[http://dx.doi.org/10.4103/1673-5374.274324] [PMID: 31997796]

[61]   Lee, J.A.; Cho, A.; Huang, E.N.; Xu, Y.; Quach, H.; Hu, J.; Wong, A.P. Gene therapy for cystic fibrosis: new tools for precision medicine. *J. Transl. Med.,* **2021**, *19*(1), 452.
[http://dx.doi.org/10.1186/s12967-021-03099-4] [PMID: 34717671]

[62]   Yiannopoulou, K.G.; Papageorgiou, S.G. Current and Future Treatments in Alzheimer Disease: An Update. *J. Cent. Nerv. Syst. Dis.,* **2020**, *12*
[http://dx.doi.org/10.1177/1179573520907397] [PMID: 32165850]

CHAPTER 7

Intellectual Properties and Current Treatment Strategies for Targeting Neurodegenerative Disorders

Abstract: Neurodegenerative disorders, including Alzheimer's, Parkinson's, Huntington's, and Amyotrophic Lateral Sclerosis (ALS), pose significant challenges due to their progressive nature and debilitating effects on patients' quality of life. Recent advancements in biomedical research have spurred the development of novel therapeutic strategies aimed at modifying disease progression and improving patient outcomes. These strategies encompass a wide array of approaches such as small molecule drugs, biologics, brain and physical exercises, gene therapy, AI-based diagnosis and treatment, nano-bioactive approaches, and innovative drug delivery systems designed to enhance Central Nervous System (CNS) penetration and target specific pathological mechanisms. Despite these promising developments, the complexity of neurodegenerative diseases necessitates multifaceted treatment paradigms that combine neuroprotection, neuronal restoration, and symptomatic management. However, several challenges persist. The Blood-Brain Barrier (BBB) remains a significant obstacle to effective drug delivery, while the heterogeneous nature of these diseases requires personalized treatment approaches. Also, safety and management of the side effects of advanced therapies such as gene therapy and brain stimulation techniques are critical concerns. The Intellectual Property (IP) landscape further complicates the development of new treatments. Navigating this complex terrain involves dealing with overlapping patents, the expiration of early patents leading to increased competition, and the high costs associated with bringing new therapies to market. Ethical and legal considerations, particularly concerning advanced technologies, add another layer of complexity.

Keywords: Blood-brain barrier, Intellectual property, Neurodegenerative disorders, Nano-bioactive approaches, Neuroprotection.

INTRODUCTION

Neurological disorders are characterized by the progressive degeneration of the structure and function of the nervous system, leading to debilitating symptoms and a profound impact on a patient's quality of life. Despite extensive research, effective treatments remain limited, primarily offering symptomatic relief rather

Shivendra Mani Tripathi, Sudhanshu Mishra, Rishabha Malviya & Smriti Ojha

than addressing the underlying causes of neurodegeneration [1]. Some recent advancements in biomedical research have spurred the development of novel therapeutic strategies aimed at modifying the treatment of disease by progression and improving patient outcomes. These strategies include a wide array of approaches such as small molecule drugs, biologics, brain exercise and physical exercise, gene therapy, AI-based diagnosis and treatment approach, nano-bioactive approach, and innovative drug delivery systems designed to enhance Central Nervous System (CNS) penetration and target specific pathological mechanisms [2, 3]. The complexity of neurodegenerative diseases necessitates multifaceted treatment paradigms that combine neuroprotection, neuronal restoration, and symptomatic management. Patents play a critical role in fostering innovation by providing exclusive rights that encourage investment in research and development. Navigating the IP landscape presents unique challenges due to the intricacies of neurodegenerative mechanisms and the need for precise and effective targeting within the CNS [4].

Various Approaches for the Treatment of NDDs

Timely identification of neurodegenerative disorders is critical to the efficacy of a given therapy in slowing down their course. Treatment can be started before the onset of severe clinical symptoms and can be considerably delayed by early detection through screening [5].

Neurodegenerative disorders are characterized by a complicated scenario of dying tissue involving various cell types and mediators, often with uncertain etiology. Regardless of where they originate, oxidative stress, neuroinflammation, and cell death are the fundamental mechanisms shared by all neurodegenerative diseases of the Central Nervous System (CNS). When compared to other therapeutic fields, the range of therapy options for neurological disease is restricted, and the medication approval rates for improved treatments continue to be low. The treatment of neurodegenerative diseases aims to slow down the degeneration's pace and enhance patients' quality of life because the degradation of neural tissue is irreversible. Pharmacotherapy alters the metabolism of neurons and glial cells to slow down degenerative processes [6].

Pharmacological Approach

One of the most difficult problems facing modern biomedicine is the development of effective therapeutic techniques for treating neurological disorders. The lack of medications that alter the pathophysiology of the disease is the main problem. Thus yet, no fundamental treatment that could provide substantial advantages to individuals suffering from harmful conditions has been created. Numerous *in vitro* and *in vivo* models are available for the research of neurodegenerative disorders, such as animal models of disease and cell models using induced pluripotent stem cells and brain organoids. Recent research has focused on microtubule stabilizing agents—natural or synthetic—that can stop tau protein disorders from destroying axons. Current therapeutic approaches can only momentarily alleviate a patient's symptoms, which include cognitive deficits and impaired motor function. On the other hand, average life expectancy has significantly grown along with the number of age-related disorders due to advances in our quality of life. Therefore, in contemporary medicinal chemistry, finding new targets for therapeutic action, creating novel synthesis techniques, and selecting possible neuroprotectors based on their targets are all crucial [7].

Since targeting transcriptional regulators would cause a broad alteration in numerous impacted downstream processes, this strategy may prove advantageous in the fight against neurodegenerative diseases [8]. This feature, however, also raises the possibility of a warning since not all genes that are involved in the pathogenesis may be negatively impacted by changes to their transcription. Abnormal protein aggregation is caused by errors in the breakdown of misfolded proteins, which is a characteristic of several neurodegenerative disorders. The inadequate elimination of pathogenic proteins, including tau, Aβ, and α-synuclein, can be ascribed to an autophagic breakdown [9].

Gene-based therapy

A potential area of research is the application of gene therapy to treat neurodegenerative disorders. This involves altering a patient's genes to halt the progression of certain diseases, such as those affecting the nervous system. There is a resurgence in gene therapy. Gene-based therapies, which encompass all types of genome alteration, hold great appeal for treating neurodegenerative disorders, for which traditional pharmaceutical techniques have shown to be mostly ineffective [10]. These therapies offer the dual promise of addressing the etiology of the disease and providing "long-term correction." A paradigm for such therapeutic intervention and a foundation are provided by the recent success of viral-vector-based gene therapy in spinal muscular atrophy, which improved survival and motor function with a single intravenous injection [11]. While there are still obstacles to overcome, the renewed hope is primarily due to advancements in the development of viral vectors that can disperse genes throughout the central nervous system and genome-engineering instruments that may modify disease pathways in previously unfeasible ways. It is impossible that spinal muscular atrophy is the only neurodegenerative condition [12].

The process of changing a person's genes to treat or prevent a disorder is known as gene therapy. Several methods are used to do this, including swapping out the faulty gene for a healthy one, deactivating the disease-causing gene, or injecting a changed gene into the organism. Research is being done on this therapy to cure viral and genetic disorders, as well as cancer [13]. Gene therapy is a promising field to research for the treatment of neurodegenerative disorders, in which most conventional medications have not been effective. Finding medications that are both safe and able to pass through the blood-brain barrier is one of the biggest obstacles in the clinical implementation of these medicines [14].

Onasemnogene abeparvovec utilizes an adeno-associated virus serotype 9 (AAV9) vector to deliver a functional copy of the SMN1 gene to motor neuron cells. The therapy bypasses the defective SMN1 gene by enabling the production of survival motor neuron (SMN) protein, which is crucial for motor neuron survival and function. The AAV9 vector efficiently crosses the blood-brain barrier after systemic intravenous administration, targeting motor neurons in the Central Nervous System (CNS) and peripheral tissues. ZOLGENSMA® is the first gene therapy approved for SMA and represents a major advancement in treating this life-threatening condition, with ongoing regulatory assessments for its intrathecal formulation in other regions [15].

Nano-bioactive Based Approach

In Neurodegenerative Diseases (NDs), neurons and their associated axons gradually deteriorate, resulting in the malfunction and eventual death of the neuronal cells. The blood-brain barrier and the blood-cerebrospinal fluid barrier are examples of natural barriers that make drug administration for the treatment of affected nerve systems infamously difficult. Nowadays, palliative care is the norm for treating a wide range of disorders. Treatment plans that focus on the disease's cause rather than its symptoms are therefore advised [16].

Drug delivery platforms based on nanotechnology present a novel means of getting around these barriers and administering drugs directly to the central nervous system, which allows for the treatment of several common neurological conditions, including Parkinson's, Alzheimer's, amyotrophic lateral sclerosis, and Huntington's. It is interesting to note that the targeting of specific mutant genes responsible for the advancement of NDs is made possible by the combination of nanomedicine and gene therapy, which may offer a much-needed boost in the fight against these diseases. Here, we went over several delivery barriers about the central nervous system before delving deeply into the most recent methods for regaining neurological function through neural stem cell differentiation [17].

A thorough overview of the history of nanomedicine's function in regulating neurogenesis through neural stem cell differentiation is provided. Many phytoconstituents are also discussed, along with their neuroprotective qualities and molecular targets in the diagnosis and treatment of NDs [18].

Stem Cell Approach

Many patients find hope in stem cell therapy, but this optimism should be balanced with the understanding that the scientific and medical communities still have a long way to go in fully understanding the complexities of stem cell biology and producing adequate data to support the logical, evidence-based application of these cells from a therapeutic standpoint. The mother or queen of all cells, stem cells are pluripotent and possess the amazing capacity to differentiate into a wide variety of bodily cell types [19]. They act as a kind of self-repairing system for the body, and as long as the human or animal is alive, they can potentially divide indefinitely to replace lost cells. A stem cell can divide and produce new cells that can either stay stem cells or develop into different types of cells with more specialized roles, such as muscle, red blood, or brain cells [20].

The non-hematopoietic group of cell progenitors known as Mesenchymal Stem Cells (MSCs), also known as mesenchymal stromal cells, are derived from the middle and outer embryonic germ layers, respectively, called mesoderm and ectoderm. MSCs are not only present in developing embryos, where they travel throughout the body during the maturation process, but they are also present, albeit in small amounts, in certain adult tissues [21]. Because MSCs possess the ability to divide and differentiate into a diverse range of specialized cell types, they are considered potential candidates for allogeneic cell therapy. Even after several decades of investigation, our knowledge of the MSC niche is still very restricted. A "niche" is a location within the body where stem cells are found along with an environment that maintains these cells in an undifferentiated form. This concept is predicated on the theory that stem cells' behaviour is determined by continual interactions with neighbouring cells [22].

Stem-cell technology presents incredible prospects. Among these include the capacity to replicate human tissues and maybe restore organ damage. The most harsh opponents of this technology are nearly silenced by its potential, but the moral complications are still very real. It is good that we have the opportunity to consider many aspects of our profession's ethics as well as our interactions with patients, businesses, and one another as we take on these problems. Before stem-cell or OEC transplantation is used on patients with neurological diseases, the experimental foundation for these procedures must be solid [23].

AI-based Diagnosis and Treatment Approach

Artificial intelligence is being used to address a variety of real-world issues, including those in the healthcare industry. It has demonstrated significant promise in hastening the comprehension and treatment of numerous medical conditions, particularly neurological ones [24]. Artificial intelligence has been applied to the medical field to analyze medical data and images, monitor patient health, diagnose diseases, find new drugs, and create new protocols.

The scientific study of the structure and cognitive processes of the brain that relate to data processing, decision-making, and interaction with the environment is known as neuroscience. Artificial intelligence is widely employed for the early identification and prediction of neurological disorders due to its capacity to analyze complex and massive volumes of data and identify patterns [25]. Understanding helps neuroscientists anticipate and identify a range of neurological conditions. In light of the foregoing viewpoint, this work represents the realization of the confluence of AI and neuroscience in the diagnosis of neurological disorders. As a result, the applications of AI for the identification and diagnosis of different neurological disorders have received particular attention [26].

AI's ability to analyse vast volumes of complex data and identify hidden patterns within it is its main strength. Because brain impulses are so complex, Artificial Intelligence (AI) is the best option for deriving conclusions and finding patterns. Using MRI data, a Support Vector Machine (SVM) could effectively identify between AD patients and those with Frontotemporal Lobar Degeneration (FTLD), as well as between AD patients and healthy persons. The D neural network design was employed to identify AD. Individuals who suffer from Moderate Cognitive Impairment (MCI) are more susceptible to AD [27].

MRI scans were analyzed using the random forest to predict the conversion of MCI to AD one to three years before clinical diagnosis. The symptoms of bacterial and viral meningitis are similar and include fever, headache, stiff neck, nausea, and vomiting [28]. It is critical to differentiate between bacterial and viral meningitis since failing to treat bacterial meningitis with the appropriate medications can result in subsequent, irreversible disorders. The subject of artificial intelligence known as Machine Learning (ML) comprises algorithms designed to identify patterns and extract significant features from vast datasets [29]. Machine learning algorithms can be used to categorize and forecast future outcomes once these patterns have been found and understood. In healthcare, ML can be used on data from various sources, assisting in diagnosis as well as disease management, tracking, and outcome prediction [30]. For example, real-time remote monitoring using Machine Learning systems can identify the severity of a disorder, document symptoms, and track a patient's reaction to therapy.

The majority of the previous methods for differential diagnosis rely on the conventional technique of calculating the area under a curve-type analysis, which allows the type of meningitis to be identified using only one predictor variable [31]. Unlike this type of approach, AI-based approaches take into account several predictor variables, such as blood albumin, blood C-reactive protein, glucose, blood soluble urokinase-type plasminogen activator receptor (suPAR), CSF lymphocytes to blood CRP ratio, neutrophil-to-lymphocyte ratio (NLR), and cerebrospinal fluid neutrophils. This results in a prediction with higher accuracy [32].

While neuro-inspired Artificial Intelligence (AI) focuses on issues where human performance surpasses that of machines, neuroscience analyses a new era of large-scale high-resolution data to find the locus of a disease or disorder for prospective therapeutic targets [33]. More training data should be made available to effectively apply data mining and Machine Learning techniques for the clinical identification of neurodegenerative disorders. To prevent a high mistake rate, additional patient data, such as postmortem data, must be available. High-accuracy outcomes, however, need the implementation of a clustering strategy for a semi-supervised algorithm [34].

Other Approaches

Brain Exercise and Physical Exercise

Exercises designed to improve thinking, memory, and attention have been widely employed in neurodegenerative disease treatment. Mnemonic strategies for episodic memory capacity training and visual information processing speed training are two examples of this type of training [35]. The moderate efficacy of cognitive training for enhancing cognitive skills was revealed by a meta-analysis of 32 studies on the subject of neurodegenerative disorders and training [36]. However, the authors advised against making firm conclusions because several of the studies that were a part of the meta-analysis had small sample sizes and were of low quality. Currently, there is little research on the efficacy of cognitive training as a therapy for neurodegenerative disorders.

One of the most frequent physiological effects of many neurodegenerative disorders is decreased levels of neurotrophic factors, primarily Brain-Derived Neurotrophic Factor (BDNF) and its receptors. Exercise and neurotrophic factor levels were found to be related to neurodegenerative disorders, according to a meta-analysis of eighteen randomized trials [37, 38]. The amount of a neurotrophic factor in blood plasma rises with physical activity. Elevated levels of neurotrophic factors mitigate the deleterious impact of nerve cell demise. Regardless of the kind of exercise, the investigations showed that increasing levels of neurotrophic factors in plasma may be achieved through aerobic, strength, or combination exercise programs [39].

Nerve/Brain Stimulation

Deep Brain Stimulation (DMS) and TMS techniques have not shown a discernible impact on clinical outcomes in neurodegenerative disorders. The DMS technique for treating Parkinson's disease motor symptoms is one exception. Patients' tremors and muscle rigidity decreased clinically significantly as a result of the DMS stimulation [40]. Nerve and brain stimulation therapies have emerged as promising approaches in the treatment of neurodegenerative disorders. Techniques such as Deep Brain Stimulation (DBS), Transcranial Magnetic Stimulation (TMS), and transcranial Direct Current Stimulation (tDCS) are at the forefront of these innovative strategies [41]. DBS involves the implantation of electrodes within specific brain regions to deliver controlled electrical impulses, modulating neural activity and alleviating symptoms in conditions like Parkinson's disease [42]. TMS uses magnetic fields to stimulate nerve cells in targeted brain areas, showing potential in treating depression and enhancing cognitive function in Alzheimer's disease. tDCS, a non-invasive method, applies a low electrical current to the scalp, influencing brain plasticity and offering therapeutic benefits for a range of neurodegenerative diseases [43]. These stimulation techniques are not only expanding the therapeutic landscape but also contributing to the development of personalized treatment approaches. The ability to tailor stimulation parameters to individual patient needs highlights the potential for improving clinical outcomes and quality of life [44].

Deep Brain Stimulation (DBS) and ablation represent critical neuromodulation techniques used in the treatment of neurological and psychiatric disorders, particularly Parkinson's disease (PD). These interventions target abnormalities in the basal ganglia-thalamocortical circuits, which are integral to motor function. Dysfunctions in this circuit propagate disruptions in downstream networks, including the thalamus, cortex, and brainstem, contributing to movement disorders. By modulating circuit activity in discrete functional domains, DBS and ablation free downstream networks to function more normally. Current evidence suggests the basal ganglia-brainstem projections also play key roles in akinesia and gait disturbances, with ongoing research exploring targets such as the pedunculopontine nucleus to address gait issues resistant to conventional therapies. DBS has become the treatment of choice for many with advanced PD and other movement disorders, offering the advantage of precise circuit modulation while minimizing disruption to uninvolved brain regions, thus overcoming the limitations of pharmacological therapies [45].

Challenges and Limitations

Advanced neurodegenerative disease treatments like gene therapy and AI-based methods present difficult moral conundrums. Despite its potential, gene therapy entails changing genetic material, which increases the possibility of unwanted genetic alterations and possibly heritable mutations. This calls into question the ethics of human DNA manipulation as well as its long-term safety [46]. Similar to this, AI-based diagnosis and treatment systems rely on enormous volumes of patient data, which, if not managed appropriately, may result in privacy violations and data exploitation. Important but difficult problems include obtaining informed consent, safeguarding patient autonomy, and preserving openness in the creation and use of these technologies. These cutting-edge treatments frequently encounter major obstacles in terms of pricing and scalability, notwithstanding their potential. Resource-intensive procedures, such as specialized equipment, highly qualified staff, and protracted research and development cycles, are required for the creation of gene therapy or nano-based medications [47]. Because of these reasons, patients, especially those in low- and middle-income nations, are unable to afford these treatments. Furthermore, underprivileged areas frequently lack the infrastructure needed to provide these therapies, such as cutting-edge medical facilities and qualified professionals. The absence of uniform regulations for new technologies such as gene therapy, nanotechnology, and artificial intelligence makes navigating the regulatory environment for advanced therapies difficult. Regulators have to strike a compromise between the requirement for comprehensive safety and efficacy assessments and the urgency of developing remedies for crippling illnesses [48]. Innovation is made even more difficult by the Intellectual Property (IP) environment. High licensing costs, overlapping patents, and patent expirations can discourage researchers or small businesses from creating novel treatments. Furthermore, the conversion of promising research into commercially viable medicines may be delayed by disagreements over IP ownership and usage rights [49]. Risks associated with advanced therapy must be carefully considered. For instance, treatments based on nanotechnology may result in nanotoxicity, which includes oxidative stress, inflammation, and even harm to healthy tissues. Nanomaterials' long-term effects on the environment are still a worry. Risks include immunological rejection, carcinogenesis, and the unpredictability of transplanted cell development present serious safety issues for stem cell treatments. Similarly, even while AI has revolutionary diagnostic potential, relying too much on algorithms without thorough clinical validation could result in incorrect diagnoses or unsuitable therapy recommendations, which would further complicate patient outcomes [50].

Patent Associated with Neurological Disorders

Several diseases that impact the nervous system are categorized as neurological disorders, including multiple sclerosis, epilepsy, Parkinson's, Alzheimer's, and others. Patents are frequently used to protect innovations in the management, diagnosis, and treatment of various diseases. The following Table **1** includes patents related to neurological disorders.

Table 1. Patents related to current treatment strategies for targeting neurodegenerative disorders.

S. No.	Patent Id	Description	Refs.
1.	The use of semaphorin-4D binding molecules for treating neurodegenerative disorders (AU2021202095B9)	This study explores methods to alleviate symptoms in subjects with neurodegenerative disorders through administering an isolated binding molecule targeting semaphorin-4D (SEMA4D) or its receptors, Plexin-B1 or Plexin-B2. The molecule, including monoclonal antibodies VX15/2503 and 67.1, inhibits SEMA4D interaction and signal transduction. Effective in conditions like Alzheimer's, Parkinson's, ALS, and Multiple Sclerosis, the treatment aims to prevent astrocyte process retraction, improving neuropsychiatric, cognitive, and motor symptoms. Specific sequences of variable heavy (VH) and light (VL) chains are identified for the antibodies. The research emphasizes the potential of these antibodies in manufacturing medicaments to address neurodegenerative and neuroinflammatory disorders, particularly in humans.	[51]
2.	Compounds and methods of use thereof for treating neurodegenerative disorders (US9539259B2)	The disclosed methods, compounds, compositions, and kits address neurodegenerative and ocular diseases by administering therapeutically effective amounts of compounds, including formulas (I), (XI), (XIIIa), (XIIIb), (XIVa), and (XIVb), notably foretinib and cabozantinib. Effective against a range of neurodegenerative conditions such as ALS, Alzheimer's, and multiple sclerosis, and ocular diseases like glaucoma and macular degeneration, these treatments promote neuronal survival, inhibit neuronal cell death, and support neurite growth. Additionally, methods include administering compounds like crizotinib, bosutinib, and dasatinib, and assessing injury through DLK and LZK protein levels, highlighting their significance in neuroprotection and nerve regeneration.	[52]
3.	Compounds, compositions and methods for protecting brain health in neurodegenerative disorders (AU2010275476B2)	The disclosed methods involve treating or preventing cognitive disorders, neurodegenerative disorders, amyloidosis-related conditions, and Alzheimer's disease by administering punicalin, punicalagin, or their derivatives to humans. These compounds can be delivered through nutraceuticals, pharmaceuticals, functional foods, dietary supplements, or botanical drugs. The treatment aims to improve or maintain cognition and memory, protect neurons, and enhance learning, attention, and reaction time. The compounds are effective in various forms, including salts, solvates, hydrates, prodrugs, enantiomers, and stereoisomers, providing a versatile approach to managing these conditions and improving overall cognitive health.	[53]
4.	Isoindoline compositions and methods of treating neurodegenerative disease (RU2692258C2)	The invention relates to isoindoline sigma-2 receptor antagonists, specifically their use in inhibiting beta-amyloid effects on neurons, which is beneficial for treating Alzheimer's disease. The compound, or its pharmaceutically acceptable salts (*e.g.*, fumarate), can be included in pharmaceutical compositions with carriers or excipients. The method involves administering an effective amount of the compound to subjects at risk of or exhibiting cognitive decline or mild cognitive impairment associated with Alzheimer's. These treatments aim to protect neurons from beta-amyloid-induced damage, thereby potentially preventing or mitigating cognitive impairments. The pharmaceutical compositions can inhibit beta-amyloid binding to neurons, thus offering a therapeutic strategy for Alzheimer's disease and related cognitive disorders.	[54]
5.	Apparatus and method for electromagnetic treatment of neurodegenerative conditions (US9433797B2)	The invention describes methods for treating neurodegenerative diseases, including Parkinson's and Alzheimer's, by using pulsed electromagnetic fields (PEMF). A PEMF device is placed external to the patient's head, generating a field with a strength of 200 milliGauss or less, which is applied through the skull to the brain or other target regions. The method aims to reduce physiological responses like inflammation and intracranial pressure, enhance neuron protection and growth, increase cytokines and growth factors, and improve blood flow. PEMF can be applied for varying durations (1 to 240 minutes) with different waveforms and frequencies. Monitoring and adjusting treatment based on physiological responses are included to optimize therapeutic outcomes.	[55]

(Table 1) cont.....

S. No.	Patent Id	Description	Refs.
6.	Exosome-based therapeutics against neurodegenerative disorders (US11369634B2)	The invention provides methods for treating demyelinating disorders by administering exosome-containing pharmaceutical compositions. These exosomes are isolated from microglia, T cells, B cells, dendritic cells, or peripheral blood mononuclear cells incubated with IFN-γ. The method involves using exosomes enriched with specific miRNAs (*e.g.*, miR-219, miR-138, miR-199a) that promote oligodendrocyte differentiation, thereby facilitating myelin growth or repair. The treatment targets disorders like multiple sclerosis, cognitive decline, Alzheimer's, Parkinson's, stroke, epilepsy, migraine, traumatic brain injury, and neuropathy. The process can include obtaining cells from the patient, incubating them with IFN-γ, harvesting exosomes, and administering them back to the patient to induce therapeutic effects. The method is particularly focused on multiple sclerosis and migraines.	[56]
7.	Apparatus and method for electromagnetic treatment of neurological injury or condition caused by a stroke (US10426967B2)	The invention describes a method for treating neurological injuries or conditions caused by stroke by reducing intracranial pressure using pulsed electromagnetic fields (PEMF). The method involves a treatment apparatus with a signal generator connected to a circular wire applicator that encircles the patient's head and a sensor to detect intracranial pressure. The applicator generates PEMF bursts of 15-40 MHz, repeating at 0.01-100 Hz. Treatment continues until intracranial pressure is reduced to a desired level, typically below 20 mmHg. Specific bursts include 2-4 msec of 27.12 MHz sinusoidal waves at 2 Hz, with variations from 1 to 10 msec and modulation of biological signaling pathways. The target region is the patient's brain.	[57]
8.	Catecholamine prodrugs for use in the treatment of Parkinson's disease (US11707476B2)	The invention provides compounds, including (4aR,10aR)-1-propyl-6-((triisopropylsilyl)-xy)-1,2,3,4,4a,5,10,10a-octahydrobenzo[g]quinolin-7-ol, as prodrugs of catecholamine for treating neurodegenerative diseases like Parkinson's. The method involves administering these compounds or their salts to patients in need. It includes various configurations of substituents on the compound. The invention extends to pharmaceutical compositions containing these compounds and methods involving their use in treating neurodegenerative or neuropsychiatric disorders. Additional agents for disease treatment can be administered alongside these compounds.	[58]
9.	For protecting the compound of brain health, composition and method in neurodegenerative disorders (CN102762573B)	The invention relates to the use of pomegranate flavonoids, specifically 4,6-(S,S)-Gallagy--D-glucose, punicalagins, or their pharmaceutically acceptable salts, in treating neurodegenerative disorders and improving brain health. These compounds are used in various forms such as pharmaceutical compositions, functional foods, dietary supplements, and medical nutrition products. The treatment aims to prevent and treat cognitive disorders, neurodegenerative diseases, and amyloidosis-related conditions like Alzheimer's. The compounds enhance or maintain cognitive functions and memory, protect neurons, and improve various aspects of memory including short-term, long-term, and procedural memory. The applications are targeted towards humans, mammals, and veterinary animals, with specific benefits in cognitive enhancement and memory improvement.	[59]
10.	Compositions and methods for ameliorating CNS inflammation, psychosis, delirium, PTSD or PTSS (US10407374B2)	The invention provides methods and compositions for treating CNS pathologies or inflammation caused by NFkB, IL-6, IL-6-R, NADPH oxidase, superoxide, and hydrogen peroxide. It focuses on using C60 malonic acid derivatives to decrease oxidative stress and inflammatory markers, offering therapeutic potential for conditions like schizophrenia, psychosis, delirium, frailty syndrome, cognitive impairments, PTSD, ALS, MS, and various dementias including Alzheimer's and Parkinson's. The derivatives can be administered *in vitro* or *in vivo*, formulated for delivery to the CNS, and used in combination with other small molecules. Additionally, the invention outlines methods for purifying C60 derivatives, ensuring their effectiveness and safety for medical applications. The methods aim to treat, slow the progression, or reverse neurodegenerative diseases and improve cognitive functions.	[60]
11.	Composition comprising vicenin-2 having a beneficial effect on neurological and/or cognitive function (CA2856599C)	The invention relates to using vicenin 2, a compound derived from various plants, for preventing, delaying, controlling, or treating neurodegenerative disorders such as Alzheimer's, Parkinson's, Huntington's, and ALS. Vicenin 2 exhibits effects like reversible acetylcholinesterase inhibition, gap junction hemichannel blockade, neuroleptic and anti-depressive effects, and TNF alpha reduction. The compound can be isolated or synthesized and used in compositions, including food products, dietary supplements, or medicaments, often in combination with other agents for enhanced neurological and cognitive benefits. Concentrations range from 0.1 μg to 500 μg, with specific formulations tailored to ensure efficacy and safety.	[61]
12.	Enhancement of delivery of lipophilic active agents across the blood-brain barrier and methods for treating central nervous system disorders (US20210145841A1)	The invention relates to edible compositions and methods to enhance the delivery of lipophilic active agents across the blood-brain barrier, resulting in higher concentrations in central nervous system tissues. The compositions include a lipophilic active agent, an edible oil, a bioavailability-enhancing agent, and an edible substrate, increasing the agent's bioavailability by at least 1.5 times. Lipophilic agents include cannabinoids, NSAIDs, vitamins, and others. Methods include contacting the substrate with an oil containing the agent and enhancing agent, then dehydrating it. The invention also covers making infused beverages from tea, coffee, or cocoa. These approaches aim to improve the treatment of central nervous system disorders.	[62]

(Table 1) cont.....

S. No.	Patent Id	Description	Refs.
13.	Herbal medicine composition for the treatment of neurological disorders and improvement of memory loss (JP5832628B2)	The invention relates to a composition containing an extract of longan meat (2.9-5.4 parts by weight), Tansan (0.1-3 parts by weight), and hemp (0.1-4 parts by weight) for preventing, ameliorating, or treating neurological diseases and psychiatric disorders. Neurological diseases include dementia, Alzheimer's, Huntington's, Parkinson's, and amyotrophic lateral sclerosis, while psychiatric disorders include depression, schizophrenia, and PTSD. The extract is a hot water extract, promoting brain functions such as learning, memory, and concentration. The composition may also include blood circulation-improving plants like Xiamen winter or gold, with detoxifying action when administered.	[63]
14.	Method of treating neurological disorders with cardiac glycosides (RU2582223C2)	The invention relates to a method for treating neurological disorders, specifically Alzheimer's disease, Huntington's disease, and stroke, using a composition containing cardiac glycosides, particularly oleandrin. The oleandrin is extracted from oleander using supercritical fluid extraction with CO_2, with or without modifiers like methanol, ethanol, or propanol. The extract includes other therapeutic agents such as oleaside A, oleandrigenin, neritaloside, and odoroside, excluding nerifoline. This method involves administering an effective dose of the composition, monitoring the clinical response, and adjusting the dosage as needed. The cardiac glycoside crosses the blood-brain barrier and remains in brain tissue for at least 8 hours. The treatment can reduce or prevent the incidence of these neurological conditions and improve the clinical status of affected individuals. The method can also be applied remotely for treating strokes, with prompt administration following stroke onset and periodic assessment of therapeutic response.	[64]
15.	Dried plant tissue and plant tissue extract for improving central nervous degenerative diseases accompanied by learning / memory disorders and movement disorders, and pharmaceuticals and foods containing them (P5906267B2)	The invention relates to a pharmaceutical product used for enhancing memory acquisition in learning/memory disorders or promoting dopamine synthesis, content, and secretion in Parkinson's disease. It contains an extract from citrus dry plant tissue (specifically Citrus reticulata blanco, Tachibana, or Giant orange) with 0.3-2.0% nobiletin, ≤0.4% narirutin, 0.1-0.8% tangeretin, and 0.4-12% hesperidin. The extract is blended with herbal ingredients at a weight ratio of 1:5 to 1:10. The product can be formulated in various forms, including granules, capsules, and tablets, and includes specific herbal ingredients such as licorice and sandalwood.	
16.	Use of ginsenoside-Rg3 in preparing medicine for preventing or/and treating dementia and medicine (AU2014352441B2)	Products for preventing and treating dementia, including Alzheimer's disease and vascular dementia. These products may be administered orally, sublingually, transdermally, intramuscularly, subcutaneously, mucocutaneously, or intravenously. Forms include tablets, capsules, powders, syrups, injection solutions, lyophilized powders for injection, creams, ointments, sprays, aerosols, gels, and patches. The content of 20(R)-ginsenoside Rg3 in the products is equal to or greater than 80%. Additionally, the medicine may contain other volatile oils from Radix Ginseng, Moschus, Rhizoma Acori Tatarinowii, Radix Curcumae, or Rhizoma Chuanxiong. The invention outlines methods for preventing and treating dementia by administering these preparations to patients.	[65]
17.	Multifunctional formulation comprised of natural ingredients and method of preparation/manufacturing thereof (EP3281614A1)	This invention relates to a multifunctional biocompatible bandage dressing for treating compromised tissues, including skin wounds, ulcers, Alzheimer's, multiple sclerosis, and melanoma. The composition includes water-solubilized nano-formulations of phyto-pharmaceuticals in a biocompatible gel or matrix, enhancing bioavailability. The bandage, made from biopolymers or synthetic polymers, delivers active ingredients like curcumin, Emblica, and Camellia extracts. It can be a patch, gauze, or film, providing a breathable, antimicrobial, and exudate-absorbing dressing. The formulation may include Aloe vera, nano-silver, and other herbal extracts, and is suitable for various applications, including diabetic foot ulcers and post-surgical wounds. It is flexible, sterilized *via* gamma irradiation, and enhances wound healing and drug delivery.	[66]
18.	Composition, formulations and methods of making and using botanicals and natural compounds for the promotion of healthy brain aging (US20220323464A1)	The disclosed compositions aim to promote healthy brain aging and prevent neurodegenerative changes leading to cognitive dysfunction. They include Huperzine A or its analog, estrogens, phytoestrogens, and vitamin D. Phytoestrogens like soy isoflavones, and vitamin D variants such as calcitriol are included. Additives like coffee and sweeteners can be added. Dosages vary, with ranges for each component specified. Compositions may include Huperzine A, soy isoflavones, and vitamin D. Pharmaceutical and nutraceutical formulations are possible, with options for immediate, extended, or timed release. They can be in tablet, capsule, powder, or patch form. Methods include promoting healthy brain aging, stimulating neurogenesis, inhibiting apoptosis, providing neuroprotection, and enhancing insulin metabolism and cerebral blood flow. Additionally, the compositions aim to modulate neurotransmitters, prevent neuronal degeneration, and alleviate symptoms of cognitive impairment and Alzheimer's disease. Monitoring methods involve assessing biomarkers and cognitive function tests to ensure efficacy.	[67]

(Table 1) cont.....

S. No.	Patent Id	Description	Refs.
19	Use of liquiritigenin in preparing medicament for treating neurodegenerative diseases (CN101152173A)	The invention pertains to a drug preparation method utilizing liquiritigenin to treat senile dementia and Parkinson's disease. Liquiritigenin, combined with excipients, is formulated into capsules, tablets, granules, or injections. It exhibits neuron protection and nourishment properties without estrogenic activity, enhancing memory and motor abilities in dementia and Parkinson's disease models. Specifically, glycyrrhizin (I) is applied in Alzheimer's and Parkinson's treatment. Compositions for treating both diseases contain glycyrrhizin (I) as the active ingredient alongside pharmaceutical excipients or carriers.	
20.	Cannabichromene (CBC), cannabidivarin (CBDV) and/or cannabidivarin acid (CBDVA) for the treatment of neurodegenerative diseases (GB2492487A)	The disclosed invention involves utilizing cannabinoids such as cannabichromene (CBC), cannabidivarin (CBDV), and cannabidivarin acid (CBDVA) for preventing or treating neurodegenerative diseases like Alzheimer's, Parkinson's, ALS, or Huntington's disease, as well as conditions like stroke, cardiac ischemia, or thromboembolism. The cannabinoids, preferably in doses ranging from 5mg to 1000mg, can be administered alone or in combination with other medicinal substances. The method encompasses administering these cannabinoids to individuals in need thereof. Specifically, the cannabinoids may be used in daily doses between 10mg and 50mg. Additionally, they can be formulated as a botanical drug substance (BDS). This invention highlights the therapeutic potential of cannabinoids in combating neurodegenerative disorders and ischemic-related diseases.	[68]

Case Studies

- A study discusses a longitudinal study on muscular strength and functional capacity decline in individuals with juvenile and adult myotonic dystrophy type 1 (DM1) over three years. The research involved 23 participants evaluated in 2016 and 2019, utilizing muscle strength testing, functional assessments, and questionnaires. Key findings include a significant decline in muscle strength, particularly in the hip abductors, knee extensors, and ankle dorsiflexors, as well as in balance (Mini-BESTest) and lower extremity functionality (LEFS). Men and those with adult-onset DM1 experienced more pronounced declines than women and juvenile-onset individuals. Physical activity, including strength training, appeared to mitigate some muscle strength losses, with differences observed across activity levels. Quantitative muscle testing emerged as a more sensitive indicator of disease progression compared to functional assessments. This study highlights the necessity of tailored clinical interventions and further research into managing DM1's impacts [69].

- A 24-year-old male patient, born on June 2, 1996, was diagnosed with Amyotrophic Lateral Sclerosis (ALS). He reported experiencing general weakness, leg cramping, and frequent falls while playing soccer. There is no known family history of ALS; however, his father has asthma, and his mother has Type II diabetes. The patient underwent a tonsillectomy at age 10. Neurological examination revealed signs of both upper and lower motor neuron involvement. Upper motor neuron findings included bilateral positive Babinski's sign and clonus. Lower motor neuron signs included hyperreflexia, with Achilles reflex graded at 4+ and patellar, biceps, and triceps reflexes graded at 3+. He has been prescribed Riluzole (100 mg orally daily) and acetaminophen as needed for symptomatic relief. Despite the progressive nature of ALS, the patient's recent outcome measures remained stable over six weeks, with no significant changes observed in key parameters. This stability was interpreted as a positive outcome, indicating a relatively consistent disease trajectory [70].

- An 18-year-old male was diagnosed with primary central nervous system lymphoma (PCNSL) after presenting with a four-month history of fever, persistent headaches, and two episodes of seizures. Diagnostic evaluation revealed tumor cells positive for leukocyte common antigen, Anaplastic Lymphoma Kinase (ALK), CD30, CD4, and other markers, confirming the diagnosis. Initial treatment included high-dose methotrexate with leucovorin rescue, intrathecal methotrexate, vincristine, procarbazine, and dexamethasone. Following the initial surgical intervention, the patient experienced no further seizures and was placed on prophylactic antiepileptic medication to prevent recurrence. The comprehensive treatment approach aims to target the tumor effectively while managing potential neurological complications [71].

- Trish Jones, a 62-year-old female, was diagnosed with Acute Motor Axonal Neuropathy (AMAN), a variant of Guillain-Barré Syndrome (GBS). She presented with bilateral motor weakness affecting her distal lower limbs, including the ankles, feet, and toes, as well as her distal upper limbs, including the hands, fingers, and wrists. Functional and mobility assessments were conducted to evaluate her condition. Trish received high-dose intravenous immunoglobulin (IVIg) therapy as needed to address the acute phase of her condition. Following 12 weeks of targeted physiotherapy sessions, she demonstrated significant improvement across key areas of her diagnosis that previously affected her independence. These improvements included enhanced muscle strength, increased range of motion, better balance, and a more stable gait, contributing to her overall recovery and functional mobility [72].

- Mr. L, a 62-year-old retired NASA astronaut, is managing Multiple System Atrophy with Parkinsonian and Cerebellar features (MSA-P and MSA-C). His symptoms include erectile dysfunction, balance difficulties, asymmetric postural tremor on the right side, bradykinesia, rigidity, urinary incontinence, and minor slurring of speech. Additionally, slight gait ataxia and balance impairments persist, raising safety concerns. Diagnostic imaging *via* T2-weighted MRI revealed bilateral hyperintense putaminal slit signs, indicative of MSA. Mr. L was initially treated with Levodopa at 250 mg four times daily for four months, followed by an increased dose of 500 mg four times daily for two months. These treatments have yielded limited improvement in his motor symptoms, reflecting the levodopa-refractory nature of MSA. He also occasionally uses CBD for pain management. Despite treatment, Mr. L continues to experience challenges related to balance and motor coordination, characteristic of the disease's progression. The persistence of gait ataxia and balance impairment underscores the need for ongoing management strategies, including physical therapy, assistive devices for mobility, and monitoring to prevent falls. His condition highlights the complex interplay of neurodegenerative and autonomic dysfunction in MSA and the critical need for tailored interventions to ensure safety and quality of life [73].

Ethical Considerations in Advanced Therapies for Neurodegenerative Disorders

While gene therapy, AI-based diagnostics, and nanotechnology-driven treatments are intriguing approaches to treating neurodegenerative diseases, they also raise serious ethical issues. Affordability and accessibility are crucial issues. Disparities in availability are caused by the high expenses of developing and providing these treatments, which frequently call for specialized infrastructure and knowledge. Accessing these therapies may be extremely difficult for patients in low- and middle-income nations, aggravating health disparities worldwide. Patients in affluent countries are also impacted by affordability, since insurance coverage for novel treatments may be restricted. The ethical ramifications of a healthcare system that could put profit ahead of fair access to treatment are called into question by this. To create structures that support affordability, such as pricing controls, subsidies, and fair license agreements, policymakers and stakeholders must work together. Careful thought must be given to these therapies' long-term effects. There are ethical concerns about manipulating the human genome and the possibility of unanticipated genetic changes and heritable consequences as a result of gene therapy. Transparency, data security, and informed permission must also be given top priority in AI-driven systems to protect patient autonomy and stop sensitive data from being misused [74].

CONCLUSION

The complicated etiology and multifaceted nature of neurological conditions, characterized by gradual nervous system degradation, provide substantial challenges. Despite an abundance of research, the majority of current therapies just provide a relief from symptoms; they do not deal with the fundamental causes of neurodegeneration. Recent developments in biomedical research have sparked the development of novel therapeutic techniques, such as gene therapy, biologics, small molecule pharmaceuticals, AI-based diagnosis and treatment modalities, and creative drug delivery systems. To enhance patient outcomes, these multimodal therapy paradigms integrate neuroprotection, neuronal repair, and symptomatic management. The patents show a promising future of more potent therapeutics that can alter the course of the disease and enhance patients' quality of life. They represent significant advances in the treatment of neurodegenerative diseases. To overcome the difficulties in developing efficient drugs for these complicated diseases, innovation must continue and intellectual property must be strategically protected.

REFERENCES

[1]   Lamptey, R.N.L.; Chaulagain, B.; Trivedi, R.; Gothwal, A.; Layek, B.; Singh, J. A Review of the Common Neurodegenerative Disorders: Current Therapeutic Approaches and the Potential Role of Nanotherapeutics. *Int. J. Mol. Sci.,* **2022**, *23*(3), 1851.
[http://dx.doi.org/10.3390/ijms23031851] [PMID: 35163773]

[2]   Upadhyay, R.K. Drug delivery systems, CNS protection, and the blood brain barrier. *BioMed Res. Int.,* **2014**, *2014*, 1-37.
[http://dx.doi.org/10.1155/2014/869269] [PMID: 25136634]

[3]   Iarkov, A.; Barreto, G.E.; Grizzell, J.A.; Echeverria, V. Strategies for the Treatment of Parkinson's Disease: Beyond Dopamine. *Front. Aging Neurosci.,* **2020**, *12*, 4.
[http://dx.doi.org/10.3389/fnagi.2020.00004] [PMID: 32076403]

[4]   Palanisamy, C.P.; Pei, J.; Alugoju, P.; Anthikapalli, N.V.A.; Jayaraman, S.; Veeraraghavan, V.P.; Gopathy, S.; Roy, J.R.; Janaki, C.S.; Thalamati, D.; Mironescu, M.; Luo, Q.; Miao, Y.; Chai, Y.; Long, Q. New strategies of neurodegeneration disease treatment with extracellular vesicles (EVs) derived from Mesenchymal Stem Cells (MSCs). *Theranostics,* **2023**, *13*(12), 4138-4165.
[http://dx.doi.org/10.7150/thno.83066] [PMID: 37554286]

[5]   Shusharina, N.; Yukhnenko, D.; Botman, S.; Sapunov, V.; Savinov, V.; Kamyshov, G.; Sayapin, D.; Voznyuk, I. Modern Methods of Diagnostics and Treatment of Neurodegenerative Diseases and Depression. *Diagnostics (Basel),* **2023**, *13*(3), 573.
[http://dx.doi.org/10.3390/diagnostics13030573] [PMID: 36766678]

[6]   Rao, A.V.; Balachandran, B. Role of oxidative stress and antioxidants in neurodegenerative diseases. *Nutr. Neurosci.,* **2002**, *5*(5), 291-309.
[http://dx.doi.org/10.1080/1028415021000033767] [PMID: 12385592]

[7]   Ballatore, C.; Brunden, K.R.; Huryn, D.M.; Trojanowski, J.Q.; Lee, V.M.Y.; Smith, A.B., III Microtubule stabilizing agents as potential treatment for Alzheimer's disease and related neurodegenerative tauopathies. *J. Med. Chem.,* **2012**, *55*(21), 8979-8996.
[http://dx.doi.org/10.1021/jm301079z] [PMID: 23020671]

[8]   Valor, L.M. Transcription, epigenetics and ameliorative strategies in Huntington's Disease: a genome-wide perspective. *Mol. Neurobiol.,* **2015**, *51*(1), 406-423.
[http://dx.doi.org/10.1007/s12035-014-8715-8] [PMID: 24788684]

[9]   Ochneva, A.; Zorkina, Y.; Abramova, O.; Pavlova, O.; Ushakova, V.; Morozova, A.; Zubkov, E.; Pavlov, K.; Gurina, O.; Chekhonin, V. Protein Misfolding and Aggregation in the Brain: Common Pathogenetic Pathways in Neurodegenerative and Mental Disorders. *Int. J. Mol. Sci.,* **2022**, *23*(22), 14498.
[http://dx.doi.org/10.3390/ijms232214498] [PMID: 36430976]

[10]   Sudhakar, V.; Richardson, R.M. Gene Therapy for Neurodegenerative Diseases. *Neurotherapeutics,* **2019**, *16*(1), 166-175.
[http://dx.doi.org/10.1007/s13311-018-00694-0] [PMID: 30542906]

[11]   Martier, R.; Konstantinova, P. Gene Therapy for Neurodegenerative Diseases: Slowing Down the Ticking Clock. *Front. Neurosci.,* **2020**, *14*, 580179.
[http://dx.doi.org/10.3389/fnins.2020.580179] [PMID: 33071748]

[12]   Puhl, D.L.; D'Amato, A.R.; Gilbert, R.J. Challenges of gene delivery to the central nervous system and the growing use of biomaterial vectors. *Brain Res. Bull.,* **2019**, *150*, 216-230.
[http://dx.doi.org/10.1016/j.brainresbull.2019.05.024] [PMID: 31173859]

[13]   El-Kenawy, A.; Benarba, B.; Neves, A.F.; de Araujo, T.G.; Tan, B.L.; Gouri, A. Gene surgery:

Potential applications for human diseases. *EXCLI J.,* **2019**, *18*, 908-930.
[http://dx.doi.org/10.17179/excli2019-1833] [PMID: 31762718]

[14] Parambi, D.G.T.; Alharbi, K.S.; Kumar, R.; Harilal, S.; Batiha, G.E.S.; Cruz-Martins, N.; Magdy, O.; Musa, A.; Panda, D.S.; Mathew, B. Gene Therapy Approach with an Emphasis on Growth Factors: Theoretical and Clinical Outcomes in Neurodegenerative Diseases. *Mol. Neurobiol.,* **2022**, *59*(1), 191-233.
[http://dx.doi.org/10.1007/s12035-021-02555-y] [PMID: 34655056]

[15] Hoy, S.M. Onasemnogene Abeparvovec: First Global Approval. *Drugs,* **2019**, *79*(11), 1255-1262.
[http://dx.doi.org/10.1007/s40265-019-01162-5] [PMID: 31270752]

[16] Nayab, D.E.; Din, F.; Ali, H.; Kausar, W.A.; Urooj, S.; Zafar, M.; Khan, I.; Shabbir, K.; Khan, G.M. Nano biomaterials based strategies for enhanced brain targeting in the treatment of neurodegenerative diseases: an up-to-date perspective. *J. Nanobiotechnology,* **2023**, *21*(1), 477.
[http://dx.doi.org/10.1186/s12951-023-02250-1] [PMID: 38087359]

[17] Tan, Q.; Zhao, S.; Xu, T.; Wang, Q.; Lan, M.; Yan, L.; Chen, X. Getting drugs to the brain: advances and prospects of organic nanoparticle delivery systems for assisting drugs to cross the blood–brain barrier. *J. Mater. Chem. B Mater. Biol. Med.,* **2022**, *10*(45), 9314-9333.
[http://dx.doi.org/10.1039/D2TB01440H] [PMID: 36349976]

[18] Zhao X. Adult Neural Stem Cells: Response to Stroke Injury and Potential for Therapeutic Applications. *Curr. Stem Cell Res. Ther.,* **2012**, *6*, 327-338.
[http://dx.doi.org/10.2174/157488811797904362]

[19] Nadig, R. Stem cell therapy - Hype or hope? A review. *J. Conserv. Dent.,* **2009**, *12*(4), 131-138.
[http://dx.doi.org/10.4103/0972-0707.58329] [PMID: 20543921]

[20] Zakrzewski, W.; Dobrzyński, M.; Szymonowicz, M.; Rybak, Z. Stem cells: past, present, and future. *Stem Cell Res. Ther.,* **2019**, *10*(1), 68.
[http://dx.doi.org/10.1186/s13287-019-1165-5] [PMID: 30808416]

[21] Kulus, M.; Sibiak, R.; Stefańska, K.; Zdun, M.; Wieczorkiewicz, M.; Piotrowska-Kempisty, H.; Jaśkowski, J.M.; Bukowska, D.; Ratajczak, K.; Zabel, M.; Mozdziak, P.; Kempisty, B. Mesenchymal stem/stromal cells derived from human and animal perinatal tissues—origins, characteristics, signaling pathways, and clinical trials. *Cells,* **2021**, *10*(12), 3278.
[http://dx.doi.org/10.3390/cells10123278] [PMID: 34943786]

[22] Lim, S.; Khoo, B. An overview of mesenchymal stem cells and their potential therapeutic benefits in cancer therapy (Review). *Oncol. Lett.,* **2021**, *22*(5), 785.
[http://dx.doi.org/10.3892/ol.2021.13046] [PMID: 34594426]

[23] Lunn, J.S.; Sakowski, S.A.; Hur, J.; Feldman, E.L. Stem cell technology for neurodegenerative diseases. *Ann. Neurol.,* **2011**, *70*(3), 353-361.
[http://dx.doi.org/10.1002/ana.22487] [PMID: 21905078]

[24] Bajwa, J.; Munir, U.; Nori, A.; Williams, B. Artificial intelligence in healthcare: transforming the practice of medicine. *Future Healthc. J.,* **2021**, *8*(2), e188-e194.
[http://dx.doi.org/10.7861/fhj.2021-0095] [PMID: 34286183]

[25] Bohr, A.; Memarzadeh, K. *The rise of artificial intelligence in healthcare applications*; Artif. Intell. Healthc, **2020**, pp. 25-60.
[http://dx.doi.org/10.1016/B978-0-12-818438-7.00002-2]

[26] Surianarayanan, C.; Lawrence, J.J.; Chelliah, P.R.; Prakash, E.; Hewage, C. Convergence of Artificial Intelligence and Neuroscience towards the Diagnosis of Neurological Disorders—A Scoping Review. *Sensors (Basel),* **2023**, *23*(6), 3062.
[http://dx.doi.org/10.3390/s23063062] [PMID: 36991773]

[27] van Leeuwen, K.G.; Schalekamp, S.; Rutten, M.J.C.M.; van Ginneken, B.; de Rooij, M. Artificial intelligence in radiology: 100 commercially available products and their scientific evidence. *Eur.*

Radiol., **2021**, *31*(6), 3797-3804.
[http://dx.doi.org/10.1007/s00330-021-07892-z] [PMID: 33856519]

[28] Yang, J.; Sui, H.; Jiao, R.; Zhang, M.; Zhao, X.; Wang, L.; Deng, W.; Liu, X. Random-Fores-
-Algorithm-Based Applications of the Basic Characteristics and Serum and Imaging Biomarkers to
Diagnose Mild Cognitive Impairment. *Curr. Alzheimer Res.,* **2022**, *19*(1), 76-83.
[http://dx.doi.org/10.2174/1567205019666220128120927] [PMID: 35088670]

[29] Mentis, A.F.A.; Garcia, I.; Jiménez, J.; Paparoupa, M.; Xirogianni, A.; Papandreou, A.; Tzanakaki, G.
Artificial intelligence in differential diagnostics of meningitis: A nationwide study. *Diagnostics
(Basel),* **2021**, *11*(4), 602.
[http://dx.doi.org/10.3390/diagnostics11040602] [PMID: 33800653]

[30] Chen, I.Y.; Pierson, E.; Rose, S.; Joshi, S.; Ferryman, K.; Ghassemi, M. Ethical Machine Learning in
Healthcare. *Annu. Rev. Biomed. Data Sci.,* **2021**, *4*(1), 123-144.
[http://dx.doi.org/10.1146/annurev-biodatasci-092820-114757] [PMID: 34396058]

[31] Liu, Q.; Cao, M.; Shao, N.; Qin, Y.; Liu, L.; Zhang, Q.; Yang, X. Development and validation of a
new model for the early diagnosis of tuberculous meningitis in adults based on simple clinical and
laboratory parameters. *BMC Infect. Dis.,* **2023**, *23*(1), 901.
[http://dx.doi.org/10.1186/s12879-023-08922-5] [PMID: 38129813]

[32] Baddal, B.; Taner, F.; Uzun Ozsahin, D. Harnessing of Artificial Intelligence for the Diagnosis and
Prevention of Hospital-Acquired Infections: A Systematic Review. *Diagnostics (Basel),* **2024**, *14*(5),
484.
[http://dx.doi.org/10.3390/diagnostics14050484] [PMID: 38472956]

[33] Chance, F.S.; Aimone, J.B.; Musuvathy, S.S.; Smith, M.R.; Vineyard, C.M.; Wang, F. Crossing the
Cleft: Communication Challenges Between Neuroscience and Artificial Intelligence. *Front. Comput.
Neurosci.,* **2020**, *14*, 39.
[http://dx.doi.org/10.3389/fncom.2020.00039] [PMID: 32477089]

[34] Khaliq, F.; Mahajani, S.; Oberhauser, J.; Wakhloo, D. Decoding degeneration: the implementation of
Machine Learning for clinical detection of neurodegenerative disorders. *Neural Regen. Res.,* **2023**,
18(6), 1235-1242.
[http://dx.doi.org/10.4103/1673-5374.355982] [PMID: 36453399]

[35] Simon, S.S.; Hampstead, B.M.; Nucci, M.P.; Ferreira, L.K.; Duran, F.L.S.; Fonseca, L.M.; Martin,
M.G.M.; Ávila, R.; Porto, F.H.G.; Brucki, S.M.D.; Martins, C.B.; Tascone, L.S.; Jr, E.A.; Busatto,
G.F.; Bottino, C.M.C. Mnemonic strategy training modulates functional connectivity at rest in mild
cognitive impairment: Results from a randomized controlled trial. *Alzheimers Dement. (N. Y.),* **2020**,
6(1), e12075.
[http://dx.doi.org/10.1002/trc2.12075] [PMID: 33204817]

[36] Bahar-Fuchs, A.; Martyr, A.; Goh, A.M.Y.; Sabates, J.; Clare, L. Cognitive training for people with
mild to moderate dementia. *Cochrane Libr.,* **2019**, *3*(3), CD013069.
[http://dx.doi.org/10.1002/14651858.CD013069.pub2] [PMID: 30909318]

[37] Ibrahim, A.M.; Chauhan, L.; Bhardwaj, A.; Sharma, A.; Fayaz, F.; Kumar, B.; Alhashmi, M.; AlHajri,
N.; Alam, M.S.; Pottoo, F.H. Brain-Derived Neurotropic Factor in Neurodegenerative Disorders.
Biomedicines, **2022**, *10*(5), 1143.
[http://dx.doi.org/10.3390/biomedicines10051143] [PMID: 35625880]

[38] Cho, S.Y.; Roh, H.T. Effects of Exercise Training on Neurotrophic Factors and Blood–Brain Barrier
Permeability in Young-Old and Old-Old Women. *Int. J. Environ. Res. Public Health,* **2022**, *19*(24),
16896.
[http://dx.doi.org/10.3390/ijerph192416896] [PMID: 36554777]

[39] Diechmann, M.D.; Campbell, E.; Coulter, E.; Paul, L.; Dalgas, U.; Hvid, L.G. Effects of exercise
training on neurotrophic factors and subsequent neuroprotection in persons with multiple sclerosis—a
systematic review and meta-analysis. *Brain Sci.,* **2021**, *11*(11), 1499.

[http://dx.doi.org/10.3390/brainsci11111499] [PMID: 34827498]

[40] Hartmann, C.J.; Fliegen, S.; Groiss, S.J.; Wojtecki, L.; Schnitzler, A. An update on best practice of deep brain stimulation in Parkinson's disease. *Ther. Adv. Neurol. Disord.,* **2019**, *12*, 1756286419838096.
[http://dx.doi.org/10.1177/1756286419838096] [PMID: 30944587]

[41] Camacho-Conde, J.A.; del Rosario Gonzalez-Bermudez, M.; Carretero-Rey, M.; Khan, Z.U. Therapeutic potential of brain stimulation techniques in the treatment of mental, psychiatric, and cognitive disorders. *CNS Neurosci. Ther.,* **2023**, *29*(1), 8-23.
[http://dx.doi.org/10.1111/cns.13971] [PMID: 36229994]

[42] Krauss, J.K.; Lipsman, N.; Aziz, T.; Boutet, A.; Brown, P.; Chang, J.W.; Davidson, B.; Grill, W.M.; Hariz, M.I.; Horn, A.; Schulder, M.; Mammis, A.; Tass, P.A.; Volkmann, J.; Lozano, A.M. Technology of deep brain stimulation: current status and future directions. *Nat. Rev. Neurol.,* **2021**, *17*(2), 75-87.
[http://dx.doi.org/10.1038/s41582-020-00426-z] [PMID: 33244188]

[43] Somaa, F.A.; de Graaf, T.A.; Sack, A.T. Transcranial Magnetic Stimulation in the Treatment of Neurological Diseases. *Front. Neurol.,* **2022**, *13*, 793253.
[http://dx.doi.org/10.3389/fneur.2022.793253] [PMID: 35669870]

[44] Carè, M.; Chiappalone, M.; Cota, V.R. Personalized strategies of neurostimulation: from static biomarkers to dynamic closed-loop assessment of neural function. *Front. Neurosci.,* **2024**, *18*, 1363128.
[http://dx.doi.org/10.3389/fnins.2024.1363128] [PMID: 38516316]

[45] DeLong, M.R.; Wichmann, T. Basal ganglia circuits as targets for neuromodulation in Parkinson disease. *JAMA Neurol.,* **2015**, *72*(11), 1354-1360.
[http://dx.doi.org/10.1001/jamaneurol.2015.2397] [PMID: 26409114]

[46] Sun, J.; Roy, S. Gene-based therapies for neurodegenerative diseases. *Nat. Neurosci.,* **2021**, *24*(3), 297-311.
[http://dx.doi.org/10.1038/s41593-020-00778-1] [PMID: 33526943]

[47] Kohn, D.B.; Chen, Y.Y.; Spencer, M.J. Successes and challenges in clinical gene therapy. *Gene Ther.,* **2023**, *30*(10-11), 738-746.
[http://dx.doi.org/10.1038/s41434-023-00390-5] [PMID: 37935854]

[48] Drago, D.; Foss-Campbell, B.; Wonnacott, K.; Barrett, D.; Ndu, A. Global regulatory progress in delivering on the promise of gene therapies for unmet medical needs. *Mol. Ther. Methods Clin. Dev.,* **2021**, *21*, 524-529.
[http://dx.doi.org/10.1016/j.omtm.2021.04.001] [PMID: 33997101]

[49] Patino, R.M. Moving research to patient applications through commercialization: understanding and evaluating the role of intellectual property. *J. Am. Assoc. Lab. Anim. Sci.,* **2010**, *49*(2), 147-154.
[PMID: 20353687]

[50] Kumah, E.A.; Fopa, R.D.; Harati, S.; Boadu, P.; Zohoori, F.V.; Pak, T. Human and environmental impacts of nanoparticles: a scoping review of the current literature. *BMC Public Health,* **2023**, *23*(1), 1059.
[http://dx.doi.org/10.1186/s12889-023-15958-4] [PMID: 37268899]

[51] AU2021202095B9 - Use of Semaphorin-4D Binding Molecules for Treating Neurodegenerative Disorders. Google Patents n.d. https://patents.google.com/patent/AU2021202095B9/en

[52] Tsai LH, Lee MS, Cruz JC. Compounds and methods for treating neurodegenerative disorders, **2003**.

[53] Leon Dupraz, P.V.; Rinsch, C. Compounds, compositions and methods for protecting brain health in neurodegenerative disorders. *Appl Filed by Amaz SA,* **2010**, *56*, 506-515.

[54] Dimitrov, M.; Alattia, J.R.; Lemmin, T.; Lehal, R.; Fligier, A.; Houacine, J.; Hussain, I.; Radtke, F.; Dal Peraro, M.; Beher, D.; Fraering, P.C. Alzheimer's disease mutations in APP but not γ-secretase

modulators affect epsilon-cleavage-dependent AICD production. *Nat. Commun.,* **2013**, *4*(1), 2246.
[http://dx.doi.org/10.1038/ncomms3246] [PMID: 23907250]

[55] US9433797B2 - Apparatus and Method for Electromagnetic Treatment of Neurodegenerative Conditions. Google Patents n.d. https://patents.google.com/patent/US9433797B2/en

[56] US11369634B2 - Exosome-Based Therapeutics Against Neurodegenerative Disorders. Google Patents n.d. https://patents.google.com/patent/US11369634B2/en

[57] A Pa, A Da, Iyer V, Diana C, Berish S. Apparatus and method for electromagnetic treatment of neurological injury or condition caused by a stroke, **2016**.

[58] JP7443606B2 - Novel Catecholamine Prodrugs for Use in the Treatment of Parkinson's Disease. Google Patents n.d. https://patents.google.com/patent/JP7443606B2/en

[59] For Protecting the Compound of Brain Health, Composition and Method in Neurodegenerative Disorders. Google Patents 2010. https://patents.google.com/patent/US20100249434A1/en

[60] Margarita Behrens, M.; Mar, D.; Ali, S.; Diego, S. *Compositions and methods for ameliorating CNS inflammation, psychosis, delirium*; PTSD or PTSS, **2018**.

[61] CA2856599C - Composition comprising vicenin-2 having a beneficial effect on neurological and/or cognitive function - *Google Patents*, n.d.

[62] JOHN D, ANDREW BC. Enhancement Of Delivery Of Lipophilic Active Agents Across The Blood-brain Barrier And Methods For Treating Central Nervous System Disorders, **2019**.

[63] Herbal medicine composition for treatment of neurological disorders and improvement of memory loss, **2012**.

[64] RU2582223C2 - Method of treating neurological disorders with cardiac glycosides - *Google Patents*, n.d.

[65] Australia I. Use of ginsenoside-Rg3 in preparing medicine for preventing or/and treating dementia and medicine, **2014**.

[66] EP3281614A1 - Multifunctional formulation comprised of natural ingredients and method of preparation/manufacturing thereof - Google Patents, n.d.

[67] Notelovitz M. Composition, formulations and methods of making and using botanicals and natural compounds for the promotion of healthy brain aging., **2014**.

[68] Roussel, M.P.; Fiset, M.M.; Gauthier, L.; Lavoie, C.; McNicoll, É.; Pouliot, L.; Gagnon, C.; Duchesne, E. Assessment of muscular strength and functional capacity in the juvenile and adult myotonic dystrophy type 1 population: a 3-year follow-up study. *J. Neurol.,* **2021**, *268*(11), 4221-4237.
[http://dx.doi.org/10.1007/s00415-021-10533-6] [PMID: 33907889]

[69] Ishibashi, K.; Ishii, D.; Yamamoto, S.; Ono, Y.; Yoshikawa, K.; Matsuda, T.; Asakawa, Y.; Kohno, Y. Decreased and Improved Movement Abilities in a Case of Myotonic Dystrophy Type 1: Examining Longitudinal Characteristics Based on Repeated Evaluations. *Cureus,* **2024**, *16*(5).
[http://dx.doi.org/10.7759/cureus.24392]

[70] Jamrozik, Z.; Gawel, M.; Szacka, K.; Bakon, L. A case report of amyotrophic lateral sclerosis in a patient with Klippel-Feil syndrome—a familial occurrence: a potential role of TGF-β signaling pathway. *Medicine (Baltimore),* **2015**, *94*(4), e441.
[http://dx.doi.org/10.1097/MD.0000000000000441] [PMID: 25634178]

[71] Ravikumar, S.; Poysophon, P.; Poblete, R.; Kim-Tenser, M. A case of acute motor axonal neuropathy mimicking brain death and review of the literature. *Front. Neurol.,* **2016**, *7*, 63.
[http://dx.doi.org/10.3389/fneur.2016.00063] [PMID: 27199887]

[72] Hafer-Macko, C.; Hsieh, S.T.; Ho, T.W.; Sheikh, K.; Cornblath, D.R.; Li, C.Y.; McKhann, G.M.; Asbury, A.K.; Griffin, J.W. Acute motor axonal neuropathy: An antibody☐mediated attack on

axolemma. *Ann. Neurol.,* **1996**, *40*(4), 635-644.
[http://dx.doi.org/10.1002/ana.410400414] [PMID: 8871584]

[73] Riley Kitchen SHTNGSB david MC. Multiple System Atrophy: A Case Study - Physiopedia. 2021.

[74] Low, J.; Ho, E. Managing ethical dilemmas in end-stage neurodegenerative diseases. *Geriatrics (Basel),* **2017**, *2*(1), 8.
[http://dx.doi.org/10.3390/geriatrics2010008] [PMID: 31011018]

SUBJECT INDEX

A

ABC score 106
Abluminal 78, 79, 80
  membrane compartments 79
  surfaces 79, 80
Abnormalities 4, 5, 6, 123
  autonomic 5
  behavioral 4
  genetic 4, 5, 6
  interventions target 123
  oculomotor 6
Acebutolol 108
Acetaldehyde 23
Acetaminophen 103, 128
Acetylcholine 21, 27, 28
Acetylcholinesterase 11
Acid 9, 23, 25, 30, 31, 49, 50, 52, 61, 84
  α-linolenic 30
  acetyl-keto-beta-boswellic 25
  amino 31
  caffeic 23
  caftaric 23
  chlorogenic 23
  cinnamic 9
  ellagic 61
  gamma-aminobutyric 23
  lactic-co-glycolic 49, 52
  linoleic 30
  oleic 30
  sialic 84
  tannic 50
Acorus calamus 26, 33
Acori tatarinowii Rhizoma 127
Activity 2, 10, 19, 27, 61, 85, 123, 128
  γ-secretase 10
  anti-tumorigenic 2
  cholinergic 27
  enzymatic 85
  estrogenic 128
  locomotor 61
  modulating circuit 123

neural 123
pharmacological 19
Acute motor axonal neuropathy (AMAN) 95, 103, 129
Adeno-associated viruses 109
Adherens junctions 79
Adiantum capillus-veneris 97
Agents 43, 67, 99, 106, 118
  anti-inflammatory 99
  cross-linking 67
  electrophysical 106
  microtubule stabilizing 118
  mucoadhesive 43
Agoraphobia 20
AI-based 116, 117, 121, 122, 124, 130
  approaches 122
  diagnosis 116, 117, 121, 124, 130
  methods 124
Aloysia citrodora 97
Alzheimer's 50, 51, 100
  disease assessment scale 100
  disease models 51
  disease treatment 50
  model rats 50
Amyloid plaques 4, 10, 25, 27
  dangerous 27
  decreased 25
  harmful 10
  massive 4
Amyotrophic lateral sclerosis 43, 44, 95, 98, 103, 116, 128
Anaplastic lymphoma kinase (ALK) 103, 129
Antioxidant 8, 11, 18, 21, 22, 24, 26, 28, 29, 31, 32, 42, 50, 67, 97
  activities 11, 26, 28
  capacity 24
  enhanced 67
  enzymes 21, 22
  polyphenolic 8
  potent 31
  properties 18, 29, 32, 50, 97
  qualities 8, 42

www.ingramcontent.com/pod-product-compliance
Lightning Source LLC
Chambersburg PA
CBHW041713210326
41598CB00007B/636